$57.35

AF796848

A Day's Adventure in Math Wonderland

A Day's Adventure in Math Wonderland

Jin Akiyama

Tokai University, Japan

Mari-Jo Ruiz

Ateneo de Manila University, Philippines

Illustrations by Frances Alcaraz

World Scientific

NEW JERSEY • LONDON • SINGAPORE • BEIJING • SHANGHAI • HONG KONG • TAIPEI • CHENNAI

Published by

World Scientific Publishing Co. Pte. Ltd.

5 Toh Tuck Link, Singapore 596224

USA office: 27 Warren Street, Suite 401-402, Hackensack, NJ 07601

UK office: 57 Shelton Street, Covent Garden, London WC2H 9HE

Library of Congress Cataloging-in-Publication Data

Akiyama, J.

A day's adventure in math wonderland / Jin Akiyama & Mari-Jo P. Ruiz.

p. cm.

Includes bibliographical references.

ISBN-13: 978-981-281-476-0 (pbk. : alk. paper)

ISBN-10: 981-281-476-0 (pbk. : alk. paper)

1. Mathematics--Study and teaching. 2. Manipulatives (Education). 3. Mathematical models. 4. Mathematics in literature. I. Ruiz, Mari-Jo P. II. Title.

QA19.M34A45 2008

510--dc22

2008009416

Book design by Irwin Cruz

British Library Cataloguing-in-Publication Data

A catalogue record for this book is available from the British Library.

Copyright © 2008 by Jin Akiyama and Mari-Jo P. Ruiz

All rights reserved.

Printed in Singapore by Mainland Press Pte Ltd

To Javy Ruiz,

may he learn to appreciate math like Ichiro

and

to Gisaku Nakamura,

may he continue to produce mathematics
into his nineties.

CONTENTS

Chapter 1 Math is Fun? 1

Chapter 2 Fat Triangles & Flattened Bagels 7

Chapter 3 Cool Curves 27

Chapter 4 A Roomful of Right Triangles 43

Chapter 5 Math in Music 59

Chapter 6 Pachinko Math 73

Chapter 7 GCF-LCM Machine 81

Chapter 8 Baumkuchen, Spaghetti & Watermelon 89

Chapter 9 Automat(h) 101

Chapter 10 A Slice of a Cone 119

Chapter 11 Paper Twists 141

Chapter 12 Fold and Cut 157

Chapter 13 Jigsaws from Tetrahedrons 171

Chapter 14 Single & Double Duty Solids 185

Chapter 15 Reversible Solids 203

Chapter 16 All the Way Home 217

Acknowledgments 225

Explanatory Notes 227

References 229

Math is Fun?

A bell signals recess time. Boys rush to the playground from all directions.

Ichiro looks around for his best friends Jai and Kino. He sees them with a group of boys gathered around Kentaro. They seem so engrossed in what he is saying.

"I wonder what Kentaro is saying that is so interesting." Ichiro moves to within hearing distance.

“It was really fun! I rode on a tricycle with square wheels ..., I ran down a musical staircase ..., I won a race on some huge slides ...,” Kentaro goes on and on breathlessly.

"Where did he go?" Ichiro asks.

"Math Wonderland," Kino answers.

"Oh, that place again." Ichiro says.

"What do you mean?" Jai wonders.

"My grandmother couldn't stop talking about that place over breakfast. She saw some TV footage and she said the kids looked like they were having fun with the mathematical models," Ichiro answers. "I can't imagine what could be fun about math," he adds.

"Math is not so bad," Jai counters.

Ichiro is doing well in school without much effort but math does not particularly interest him. He finds his math class just a notch above boring. At times, he can hardly keep awake. Besides, his math homework takes much time away from computer games, TV, and his other favorite activity – taking apart and putting together his toy robots.

Kino, always eager and impulsive, taps him on the shoulder, "Let's go there this weekend."

"I'd like to see it," says Jai.

On his own, Ichiro would not have planned on going, but with his best friends, it might be an adventure.

The next weekend, Ichiro, Jai and Kino take the bus and soon they find themselves standing in front of an ordinary two-story building.

"Are we at the right place?" Kino wonders.

He gets to the door first, with Ichiro and Jai trailing him. They all see a sign above:

MATH WONDERLAND

They slowly push the door open, wondering whether the visit will be a waste of time.

As soon as they get inside, they hear the shrill voices of many excited kids.

Fat Triangles & Flattened Bagels

Keiko

At the door, the boys are greeted by a young woman. Her name tag says Keiko. The first thing they notice are her roller skates.

"Strange wheels," Kino comments, "like bagels flattened along the edges."

"There's a fat triangle shape turning inside the square at the center." Ichiro adds.

"And the fat triangle touches every part of the square as the wheels turn," Jai adds as he observes the wheels.

"Cool, I've never seen wheels like that before," Kino says wondering where he can buy the skates for himself.

But before he can even ask, Keiko says, "You can borrow skates like these for going around Wonderland."

She takes them to a counter where they get skates. They are very eager to go to the exhibition rooms. The skates move them smoothly across the floor.

"How do these things work?" Ichiro asks.

"You'll see," Keiko answers as she leads them to a ramp going to the second floor.

There they find a room where there are more objects that look like fat triangles and flattened bagels. The sign beside the door says:

CURVES OF CONSTANT WIDTH

In one corner is a miniature manhole and a cover both in fat triangle shapes. There are other sets of manholes and covers too – round, square, triangular, trapezoidal.

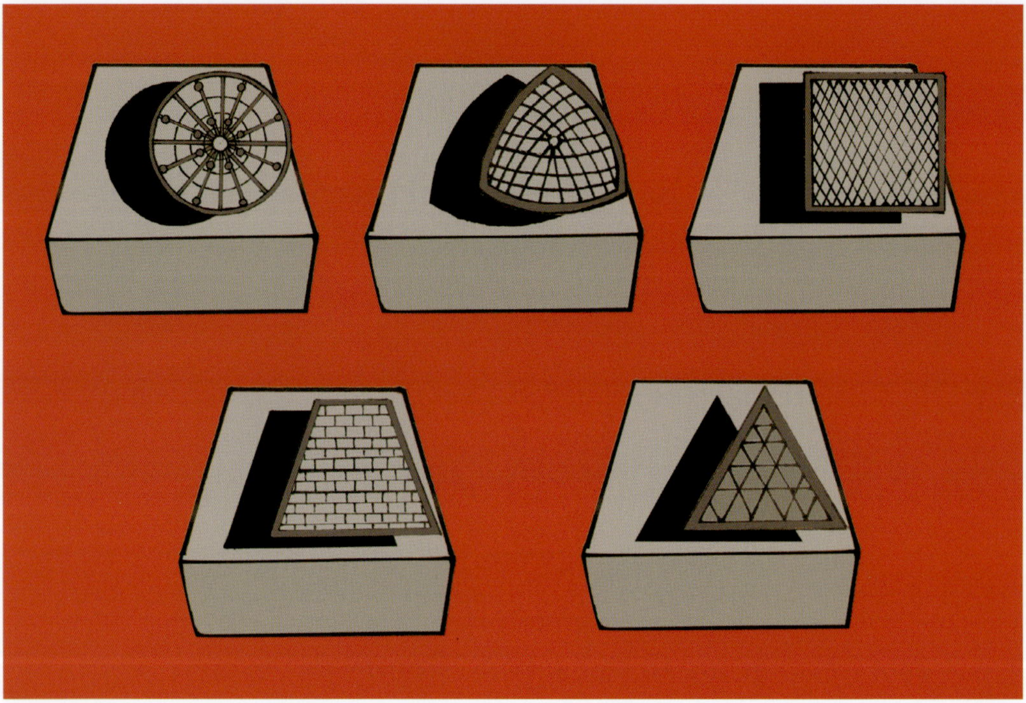

Keiko leaves them with one of the guides in the room.

"Call me Koji," he says. "Why don't you move the covers around and see what happens?" he suggests.

They each handle one shape and then another. Jai chooses the square manhole and cover.

"Hey, look, the cover falls in," he calls out to his friends.

"Do you know why?" Koji asks.

"Of course! The side is shorter than the diagonal, so when I move the cover this way, it falls in."

Among the three friends, Jai has the more analytical mind. His knowledge is both wide and deep.

Ichiro and Kino are busy making their own discoveries.

"The triangle and trapezoid also fall in, but not the circle and the fat triangle," Ichiro observes.

"What's so special about those two shapes?" Kino inquires.

"They are bounded by curves of constant width" Koji answers.

He shows them posters that explain and illustrate the concept.

Curves of Constant Width

The width of a curve is measured by the perpendicular distance between two opposite parallel lines tangent to its boundary. If the distance is the same regardless of the direction of the tangent lines, then the curve is said to have constant width.

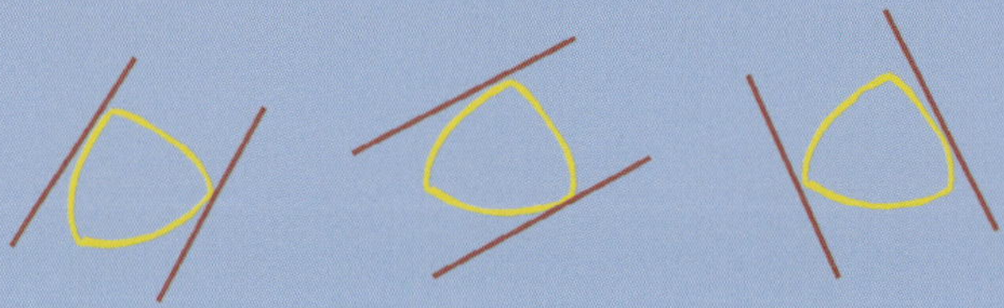

Construction of a Reuleaux Triangle

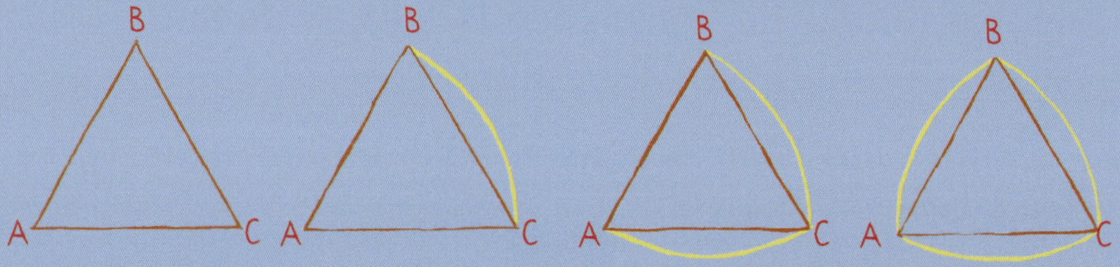

Start with an equilateral triangle ABC.

Draw the arc BC of the circle centered at A, the arc CA of the circle centered at B and the arc AB of the circle centered at C.

Since ΔABC is equilateral, all these circles have the same radius, the arcs BC, CA and AB comprise the boundaries of the Reuleaux triangle.

"Reuleaux triangle – so that's what the fat triangle is called," Kino observes.

Construction of Other Curves of Constant Width

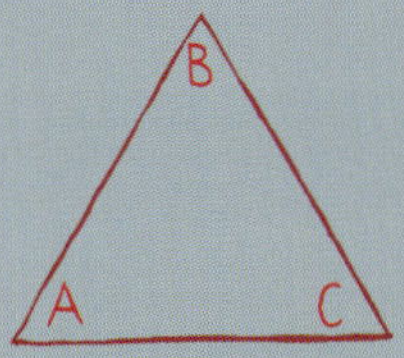

Start with an equilateral triangle ABC.

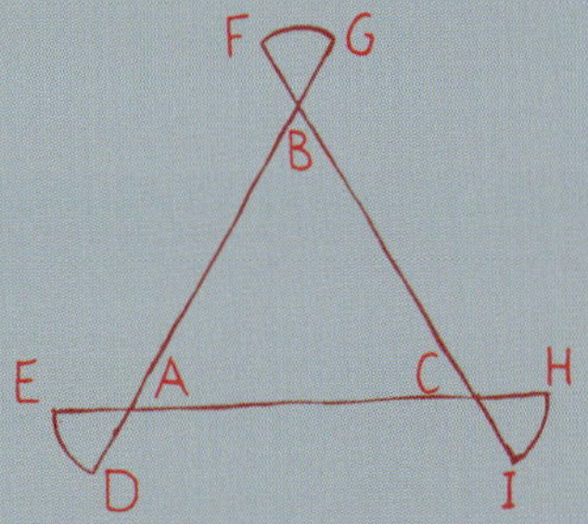

At each vertex, extend the intersecting sides the same distance beyond the vertex. This will create 60° angles external to the triangle.

Draw the arc DE of the circle centered at A, the arc FG of the circle centered at B, and the arc HI of the circle centered at C. Since the sides of the triangle were extended the same distance beyond its vertices, all these circles have the same radius.

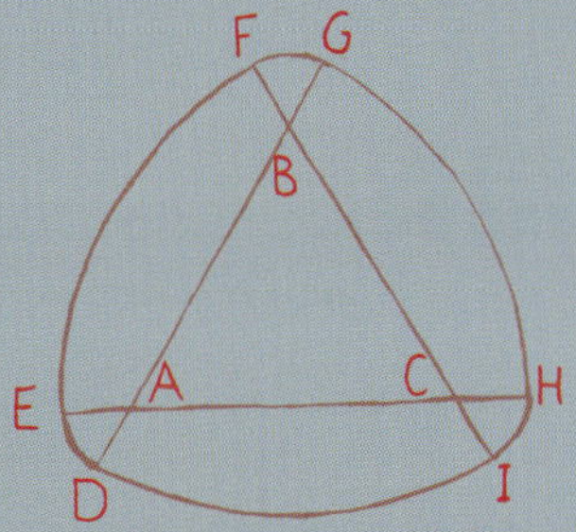

Draw the arc FE of the circle centered at C, the arc GH of the circle centered at A and the arc ID of the circle centered at B. All these circles have the same radius. All the arcs together comprise the boundary of a curve of constant width.

As a demonstration, Koji rolls several shapes between two parallel lines and shows that the circle and Reuleaux triangle touch both lines all the time as they roll along, while the square does not.

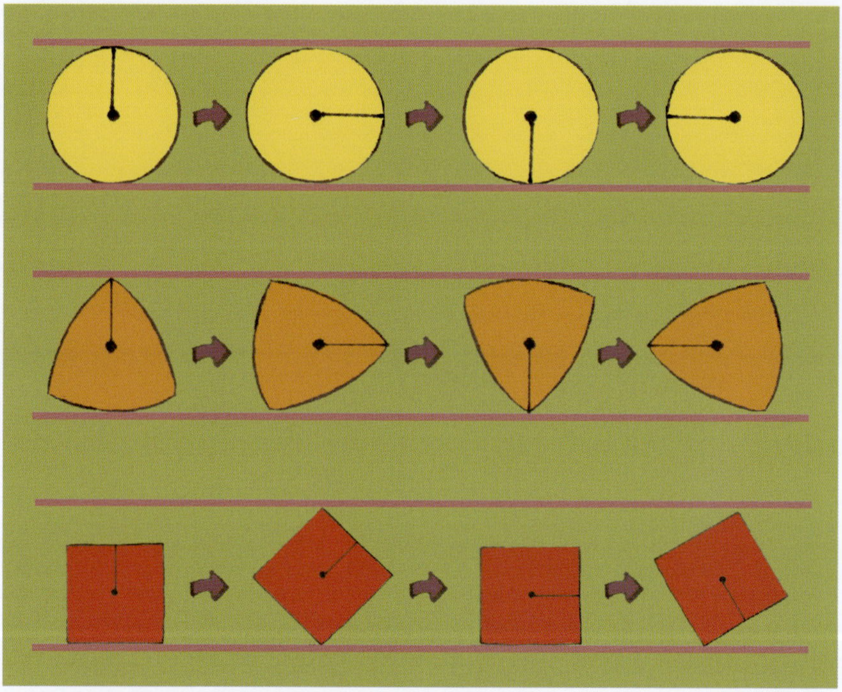

“Shapes of constant width will roll smoothly along a flat surface. Don’t you feel how smoothly your skates move?”

“Are there other curves of constant width?” Kino asks.

“The Reuleaux concept extends to pentagons, heptagons and so on,” Koji answers as he shows them a frame containing a coin from Bermuda shaped like a Reuleaux triangle, and an old British coin shaped like a Reuleaux heptagon.

In another part of the room is a sign that says:

SQUARE HOLE

"This remarkable machine makes square holes," Koji announces.

"Really?" Ichiro remarks.

Jai and Kino are also skeptical.

"Here, one of you hold this piece of foam against the blade and test it," Koji says.

Ichiro steps forward and peers at the machine. He is very interested in mechanical things. The blades form parts of a streamlined fat triangle. When Koji switches the machine on, Ichiro observes that the blades move within a square. Their movement is from the top left to right, then down the right side, and when they reach the bottom, they move from right to left and then back to the top, just like the movement of the fat triangle in the skates. As they move, they touch every point on the boundary of the square except for the very corner points.

"Go on, place the foam against the blade," Koji instructs.

Ichiro does as he was told and when the machine is switched off, there is indeed a square hole on the piece of foam, although it is slightly rounded at the corners.

"Neat!" the boys chorus.

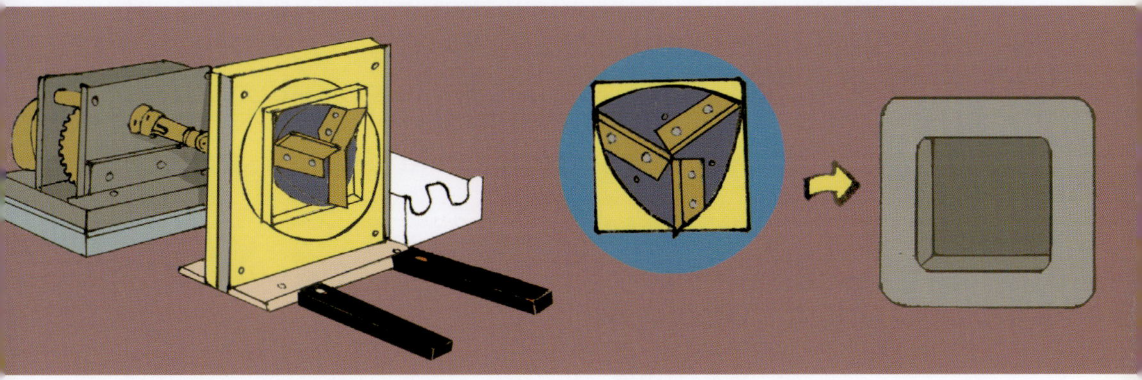

The sign beside the next machine says

HEXAGONAL HOLE

They see that the blades look like parts of a streamlined fat pentagon. They are no longer skeptical. They believe that they will see a hexagonal hole. Kino takes part in the demonstration merely to confirm it.

But Jai is thinking way ahead of everyone else. He has a eureka moment.

"Fat triangle blade, square hole, fat pentagon blade, hexagonal hole. It figures! It will take a fat heptagon blade to make an octagonal hole and so on and so on," he says triumphantly.

"That is exactly the case," Koji says in encouragement.

"Can you make holes with an odd number of sides?" Ichiro asks.

"Yes we have blades that make triangular holes and pentagonal holes but they are not of constant width," Koji replies.

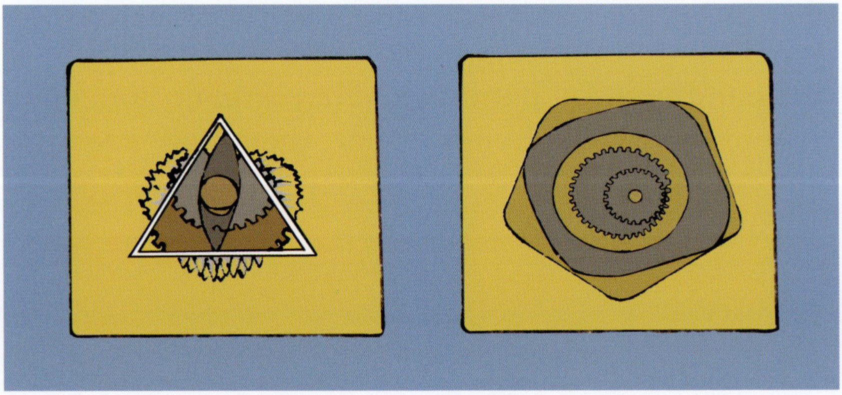

They make a detour and he shows them the blades.

"But there's more," Koji announces.

He leads them to another machine and says "This is a rotary combustion engine you find in cars."

"There's that fat triangle again – inside a capsule," Kino observes.

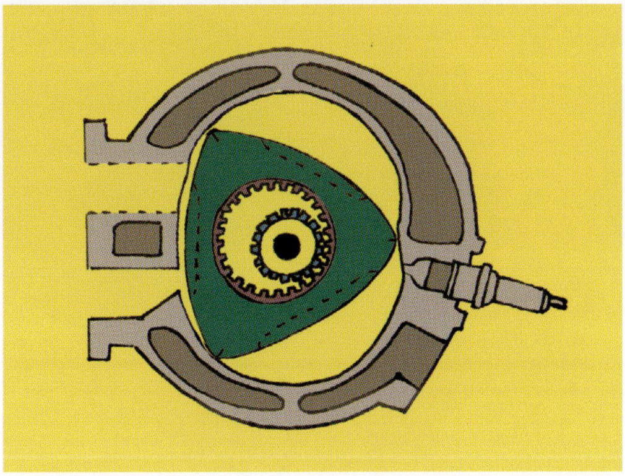

"Right, the capsule is called the bore of the engine and that Reuleaux triangle inside the bore is a rotor," Koji continues, "The shape of the capsule is based on an epitrochoid curve. This curve is formed by tracing the midpoint of a radius of a circle as it moves around the circumference of another circle with twice its diameter."

A poster on a wall shows various epitrochoid curves. Koji points it out.

Epitrochoid

An epitrochoid curve is a curve traced by a point P attached to a circle of radius b rolling around on the circumference of a fixed circle of radius a.

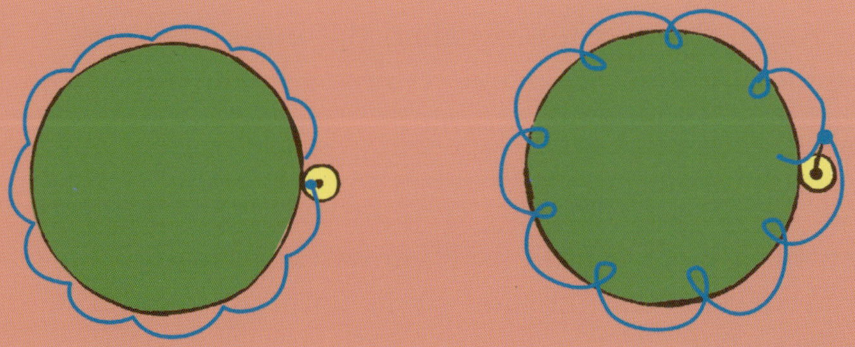

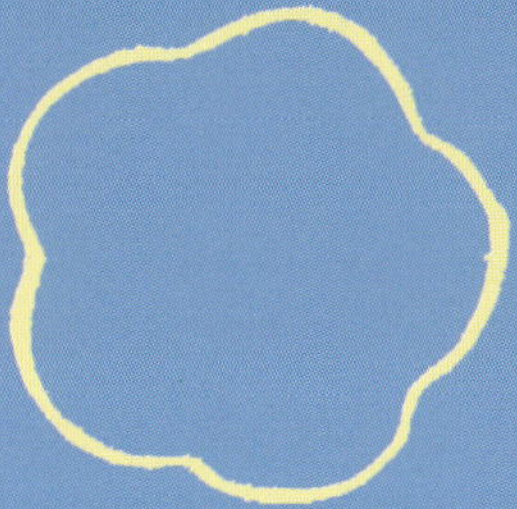

$a=1,\ b=.2,\ h=.1$

$a=1,\ b=.4,\ h=1.2$

$a=1,\ b=.6,\ h=.6$

$a=1,\ b=.8,\ h=2.6$

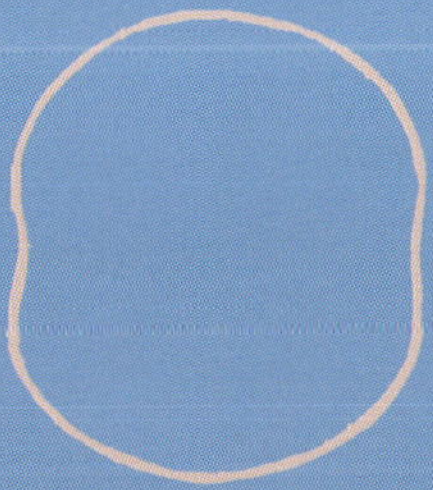

$a=1,\ b=.5,\ h=.25$

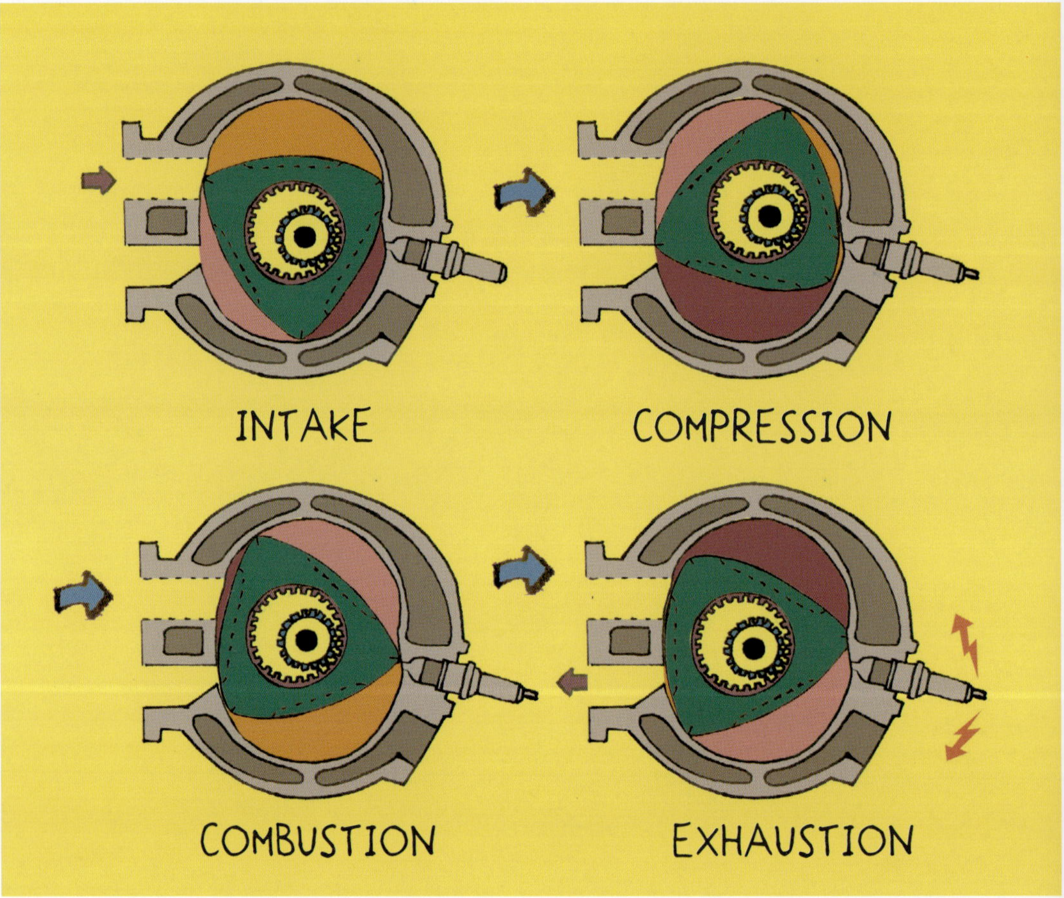
INTAKE
COMPRESSION
COMBUSTION
EXHAUSTION

Koji demonstrates. "See, the three vertices of the rotor touch the bore at three points creating three chambers. As the rotor turns, each of the chambers alternately expands and contracts."

He continues, "The expansion of the intake chamber draws in a mixture of fuel and air. Its contraction compresses the mixture and moves it towards the spark plug. The spark plug ignites the fuel. As the fuel burns, the gases expand and push the rotor. This causes the exhaust to be expelled. So the process involves intake, compression, combustion, exhaust."

"Combustion creates the power that makes the car move," he concludes.

Ichiro did not completely understand this but he is convinced of the usefulness of the Reuleaux triangle and so are his two friends.

They thank Koji and move on, eager to see and experience more.

Cool Curves

The boys are drawn to the voices of excited kids who are cheering someone on.

"Let's see what is going on," Ichiro suggests.

"Sounds like fun," Kino agrees.

As they enter the space, they see four huge slides. Three have curved shapes and the fourth is straight. Four kids are poised for a race down.

"I think the one on the straight slide will win," Kino ventures. "Didn't we learn that the shortest distance between two points is a straight line?"

Ichiro and Jai are not so sure.

"Let's just watch first," Jai cautions.

A guide signals the start of the race. Off they go. It's the kid on the second slide from the left who wins. As soon as the first group lands, another group of four are getting ready to race down.

They watch again, and again they see that the kid on the same second slide wins.

"Is that just a coincidence?" Ichiro wonders.

They watch a third race with the same result.

"Why?" they ask each other.

The guide running the race is Miki. He is calling for volunteers. The three friends rush to volunteer. One other boy is chosen to join them. They take off their skates and Miki weighs each boy. He gives them sandbags to stuff in their pockets to equalize their weights and something to sit on to make the slide down smooth. All three want to get on the second slide but Miki assigns it to Ichiro.

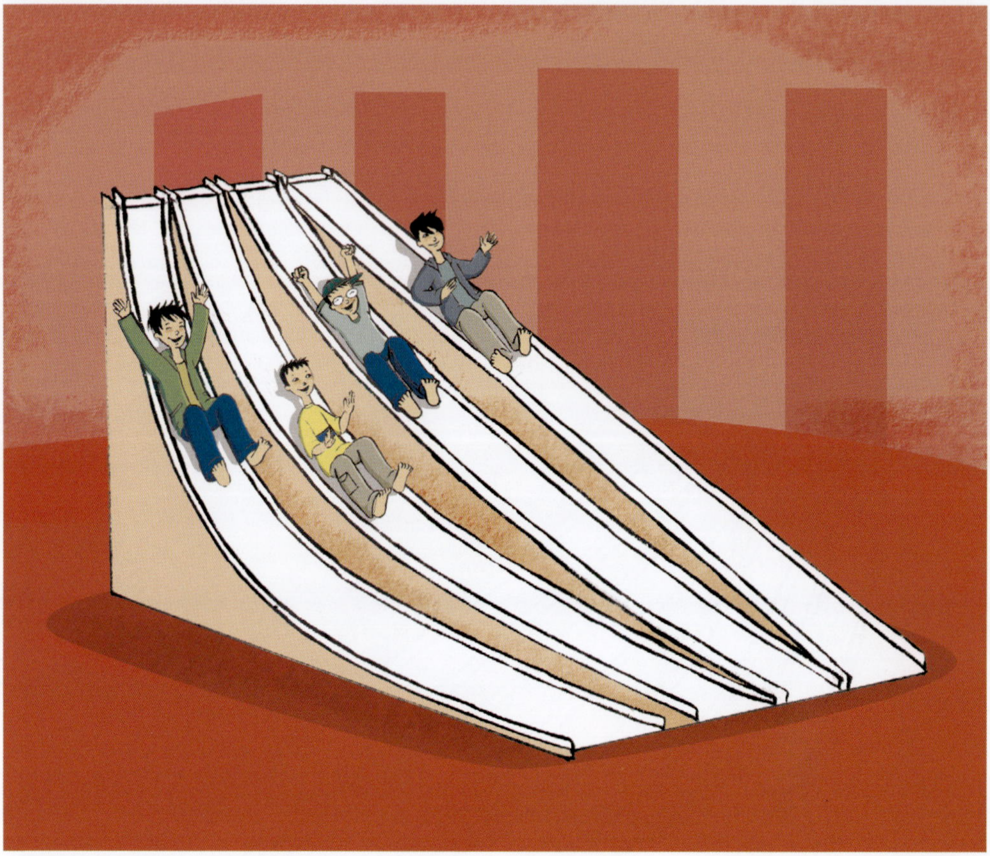

As expected, Ichiro wins. They approach Miki and ask, “Can you tell us why?”

“That curve on the second slide from the left is very special. It is called a cycloid.”

He leads them to a poster on the wall.

Cycloid

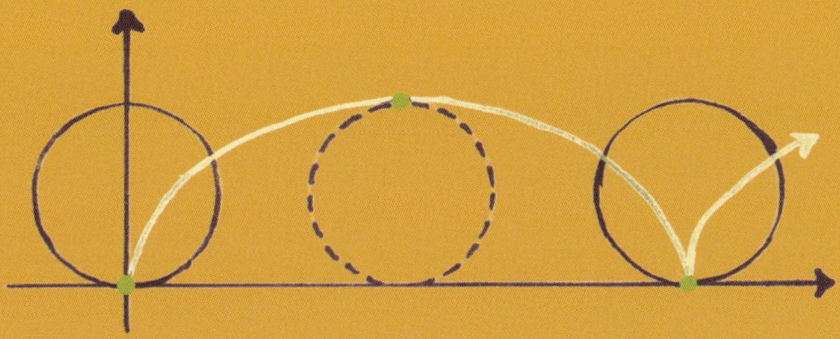

A cycloid is a curve traced by a fixed point on the circumference of a circle, as the circle rolls along a straight line.

Path of Most Rapid Descent

Among the infinitely many curves which join two given points, one lower than the other, the inverted cycloid is the curve which is traversed in the shortest time by a particle moving from the higher point to the lower point, under the pull of gravity. This property was discovered by the Swiss mathematician, Johann Bernoulli, in 1669. It is called the brachistochrone property.

Tautochrone Property

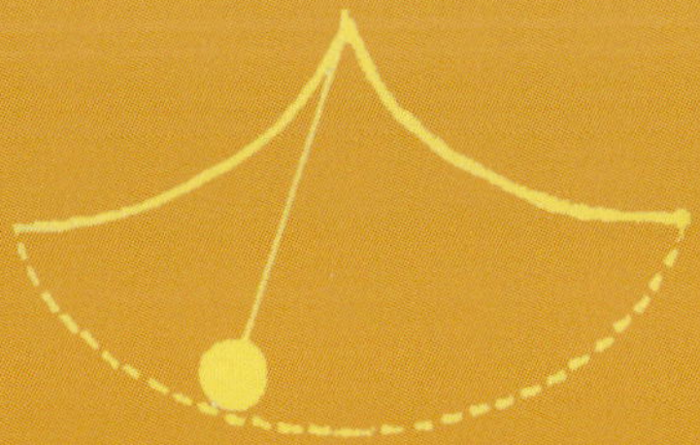

The inverted cycloid also has the property that the time it takes for the particle to reach the lower point is the same regardless of the point on the inverted cycloid at which the particle is release. A curve with this property is called a tautochrone.

This property was discovered by the Dutch scientist, Christian Huygens, in1673. He used it in the design of the first pendulum clock. The pendulum is forced to swing in an inverted cycloid arc, bounded above by two half arcs of the same inverted cycloid, so that it will swing in approximately equal time regardless of the size of the swing.

"Can you explain that further?"

Jai is referring to a statement on the poster stating that the cycloid has the special property that the time it takes for a particle to reach the lower point is the same regardless of the point at which the particle is released.

"The explanation involves some differential equations, let me just show you." Miki offers.

They head back to the slides.

"OK, one of you climb up to the second slide and slide down."

Kino is quick to volunteer.

"I will time you."

Kino clocks in at 1.34 seconds.

"Now slide down again, but this time start three fourths of the way up."

Kino does as he was told and clocks in again at 1.34 seconds.

"That's amazing," Ichiro says.

Jai is thinking the same thing.

"Try it again starting at half way down," Jai suggests.

The trial again results in a time of 1.34 seconds. Jai makes a mental note: tautochrone. Kino is happy to be the guinea pig.

As they thank Miki, they notice that in another part of the room, kids are lining up for something. Kino goes ahead to investigate and returns quickly.

“It’s the tricycle with the square wheels, the one Kentaro was talking about.”

Kino gets in line, the other two decide to come closer and have a look.

“It’s really moving forward, but look at the road it’s traveling on. It’s a curved road,” Ichiro observes.

“Maybe that’s another special curve,” Jai ventures. “I wonder if it can travel on other curved roads.”

One of the guides, Hiro, overhears them and tries to explain.

“There are some conditions that must be met so that the square wheel can move smoothly over a curved road. The length of a side of the wheel must be equal to the length of a segment of the curve. As the wheel moves forward, its side is always tangent to the curve. Also the center of the wheel should be moving in a straight line.”

Hiro leads them to a computer screen where they see a square moving forward along a curved road.

"The road bed consists of inverted catenaries placed end to end," Hiro says.

"What's a catenary?" Ichiro inquires.

"It's the kind of curve formed when a rope hangs loosely between two supports," Hiro answers.

On the computer screen they see other polygons rolling along roads consisting of inverted catenaries.

"Look, pentagons and hexagons will also work." Ichiro points out.

"But the catenaries are getting flatter," Jai observes.

"Does it work with regular polygons that have more sides?" Ichiro asks.

"Yes, but as the sides increase, the regular polygon becomes more and more like a circle and the catenaries become more and more like a straight line," Hiro explains.

Jai is quietly absorbing the fact that as the number of sides of a regular polygon increases, the shape of the polygon approaches a circle.

"I have a problem for you to think about," Hiro announces. "It won't work for an equilateral triangle. Figure out why!"

Jai is intrigued and makes a mental note of the problem.

Kino is back from his tricycle ride. He pats Ichiro on the shoulder.

"Fun ride!" he says. "But it takes some effort to get the tricycle to move forward."

On a poster is a clothoid, another curve they have never seen before.

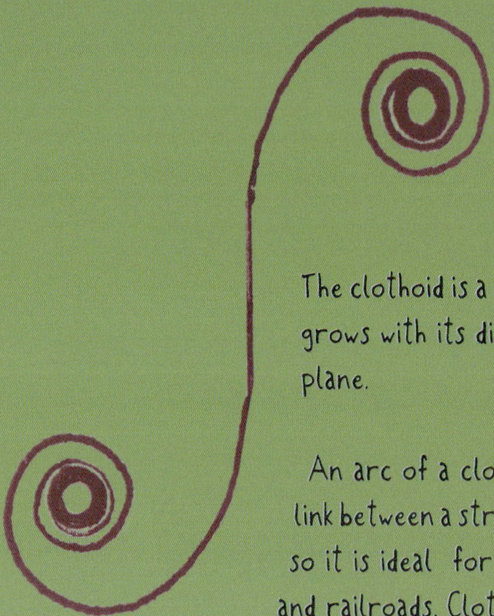

Clothoid

The clothoid is a double spiral whose curvature grows with its distance from the origin of the plane.

An arc of a clothoid provides the smoothest link between a straight line and a circular curve, so it is ideal for use in the design of highways and railroads. Clothoid arcs are also used in the design of movements of robot vehicles.

Roller coasters today have clothoid loops rather than circular loops of earlier designs. With circular loops, the train slows down at the top so greater entry speeds are needed to complete the loop. These cause the trains to go too fast at the end of the loop. A clothoid loop with a smaller radius at the top and a large radius at the bottom solves the problem.

CLOTHOID LOOP

CIRCULAR LOOP

1. A roller coaster in which the riders go through fourteen inversions (three track and eleven seat inversions) found in the Fuji-Q Highlands amusement park in Japan, built at the cost of over US$31M.

The part about the movements of robot vehicles catches Ichiro's attention. Kino is recalling the *eejanaika*[1] roller coaster ride they took in Fuji-Q Highlands.

"The *eejanaika*, were those clothoids or circular arcs?" he asks.

"Clothoids, I think," Ichiro answers.

Jai asks, "Why are clothoid loops safer?"

Hiro tries to explain, "Its physics, the interaction between speed, centripetal acceleration and gravity."

Since physics is still alien territory for the boys, they leave it at that.

A Roomful of Right Triangles

"All is number"
Pythagoras
(c. 580-500B.C)

A line has formed in front of one of the rooms. A sign says

PYTHAGORAS ROOM

"What's a pythagoras?" Kino asks.

"It's not a what, it's a who," Ichiro counters.

"He's a Greek mathematician from way back," Jai clarifies.

"He's probably very important to have a room named after him," Kino muses.

As they wait in line, they can see inside the room. It is awash in color. Brightly colored walls make the ivory floor stand out.

"Look at the floor, its tiled with triangles," Ichiro observes.

"Isosceles right triangles," Jai says.

As they enter the room, they see a sketch of Pythagoras occupying a prominent place on a wall.

"All is number"

Pythagoras

(c. 580-500 B.C.)

Pythagoras was one of the most influential thinkers of the sixth century B.C. He founded a school in Crotona, Southern Italy, where his followers studied number theory, geometry, music and astronomy. Books 7, 8 and 9 of Euclid's thirteen volume work "Elements" are devoted to Pythagorean number theory.

"All is number, what does that mean?" Kino wonders.

A guide, Miho, overhears him. She approaches and says, "Pythagoras believed that everything in the universe is connected to numbers. So he and his followers thought they could uncover the secrets of the universe by studying the properties of numbers."

On one side of the room are several rotating contraptions. Many kids are gathered around them. Kino squeezes his way to the front. Ichiro and Jai slowly inch their way in.

The first device has a right triangle in the center and on each side of the triangle is a square.

As the device turns, colored liquid from the two smaller squares flow into the largest square.

The smaller squares empty completely as the largest square is filled. A label reads

PYTHAGOREAN THEOREM

$$x^2 + y^2 = z^2.$$

"What does that equation have to do with the triangle and squares?" Kino asks.

Ichiro takes some time to think it out. Finally, he says "x and y stand for the lengths of the legs of the triangle, z for the length of the hypotenuse."

"x^2 and y^2 are the areas of the squares on top of the legs, z^2 is the area of the square on the hypotenuse," Jai continues, "so the equation says that the areas of the squares on the legs, when added together, equal the area of the square on the hypotenuse. For example, if

$$x=3,\ y=4 \text{ and } z=5, \text{ then } 9 + 16 = 25."$$

"That's why the liquid fills up the largest square in the device when the other two are empty," Ichiro adds.

"OK, I get it," Kino replies.

"The Pythagorean Theorem says that this is true for any right triangle," Miho tells them.

Jai makes another mental note to check the formula out on various right triangles when he gets home. Ichiro is really more interested in the mechanism that makes the device turn and how the colored liquid flows in and out. He stands in front of the device for a long time. Kino is already moving on.

In the second device, the same thing happens except that in place of the liquid, slivers of colored plastic move from the two smaller squares to fill up the largest square.

"How are the squares broken up into pieces?" Jai asks.

Miho leads them to another poster which explains the dissection.

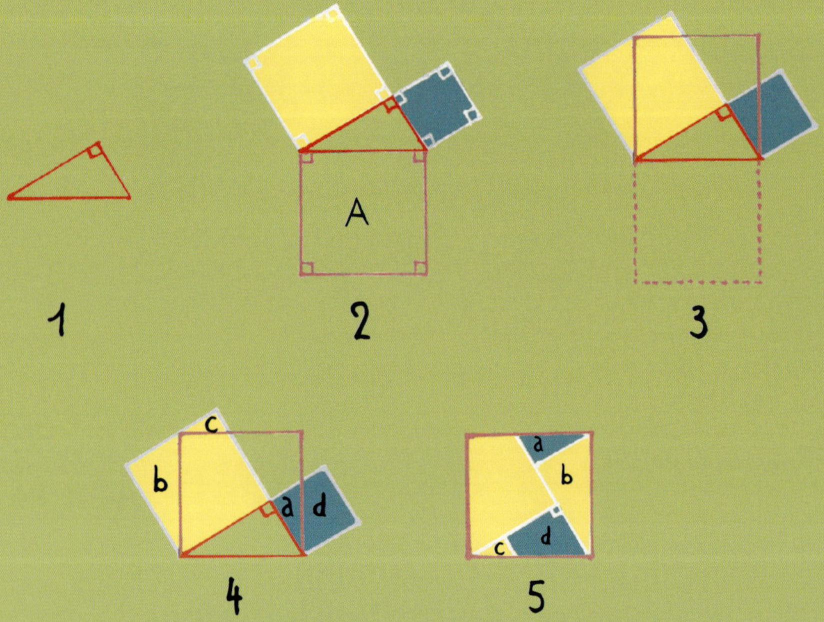
A
1
2
3
c
b
a
d
4
a
b
c
d
5

"I wonder how Pythagoras ever thought of that," Ichiro says.

"There's a story that says Pythagoras got the idea by observing the tiles on the floor of a temple. The tiling pattern was like the one on our floor here. Can you find the Pythagorean Theorem among the tiles?"

The boys take a careful look. They see the pattern emerge. Ichiro traces it on the floor.

"Hey, the other devices don't have squares on the right triangles," Kino observes.

When the other two turn to look, they see semicircles, pentagons and even elephants!

"The theorem can be extended, the area relationships still work when the squares are replaced by other similar figures," Miho explains.

"How does it work?" Jai pries.

Miho leads them to a board to explain:

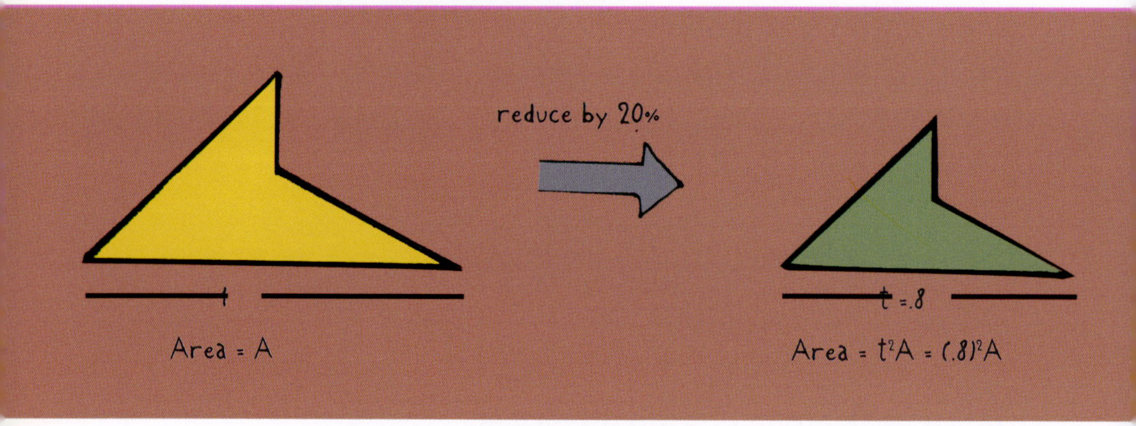

"Have you observed how a copying machine reduces an image?" she asks.

"Suppose I have some figure with area A and I want a copying machine to reduce it by 20." She draws on the board.

"Then the resulting figure will be similar to the original figure and its area is t^2A or $(.8)^2A$, she continues. "It will work in a similar way if I want to enlarge the figure."

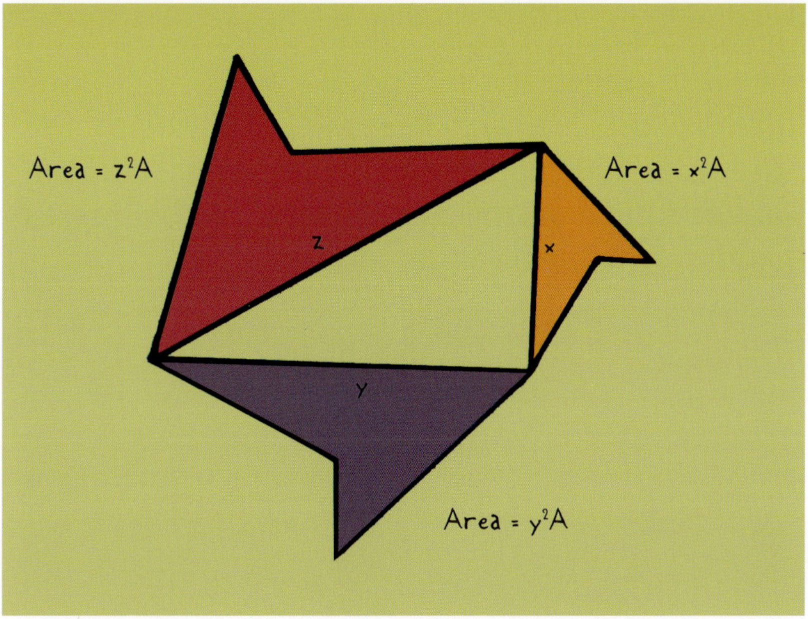

She continues to draw on the board.

"Let me put similar figures on the sides of the triangles and scale their areas to the lengths of the sides," she says. "From the Pythagorean Theorem, we know

$$x^2 + y^2 = z^2$$

so

$$\begin{aligned} x^2A + y^2A &= (x^2 + y^2)A \\ &= z^2A." \end{aligned}$$

"Whatever the shapes of the figures on the triangle, as long as they are similar, and the areas proportional to the sides, then the sum of the areas of the figures on the legs equal the area of the figure on the hypotenuse," Jai expresses concisely while nodding his head.

"That's it!" Miho says.

They soon see another wooden model. It has two toy cars, one moving up and down an incline and the other up and down a helical ramp winding inside a cylinder.

"What is that about?" Kino asks.

“That shows you how to get the length of a helix,” Miho explains. “Watch the cars. They start down at the same time -- one from the top of the incline and the other from the top of the ramp, and they move at the same speed. The gears trigger their descent.”

“They get down at the same time. That means the length of the helix is the same as the length of the incline,” Ichiro concludes.

“Precisely,” Miho confirms.

Ichiro carefully observes the movement of the gears that cause the cars to move up and down.

“But it’s here in the Pythagoras Room, what’s it got to do with Pythagoras?” Kino inquires.

“Think of unwinding the sections of the helix and flattening them,” Miho says.

“Let’s call the height of the cylinder h, and its base r. Now suppose that the helix winds around the cylinder n times -- in the model $n = 4$,” she continues.

She draws on a board.

"The width of each flattened section is the

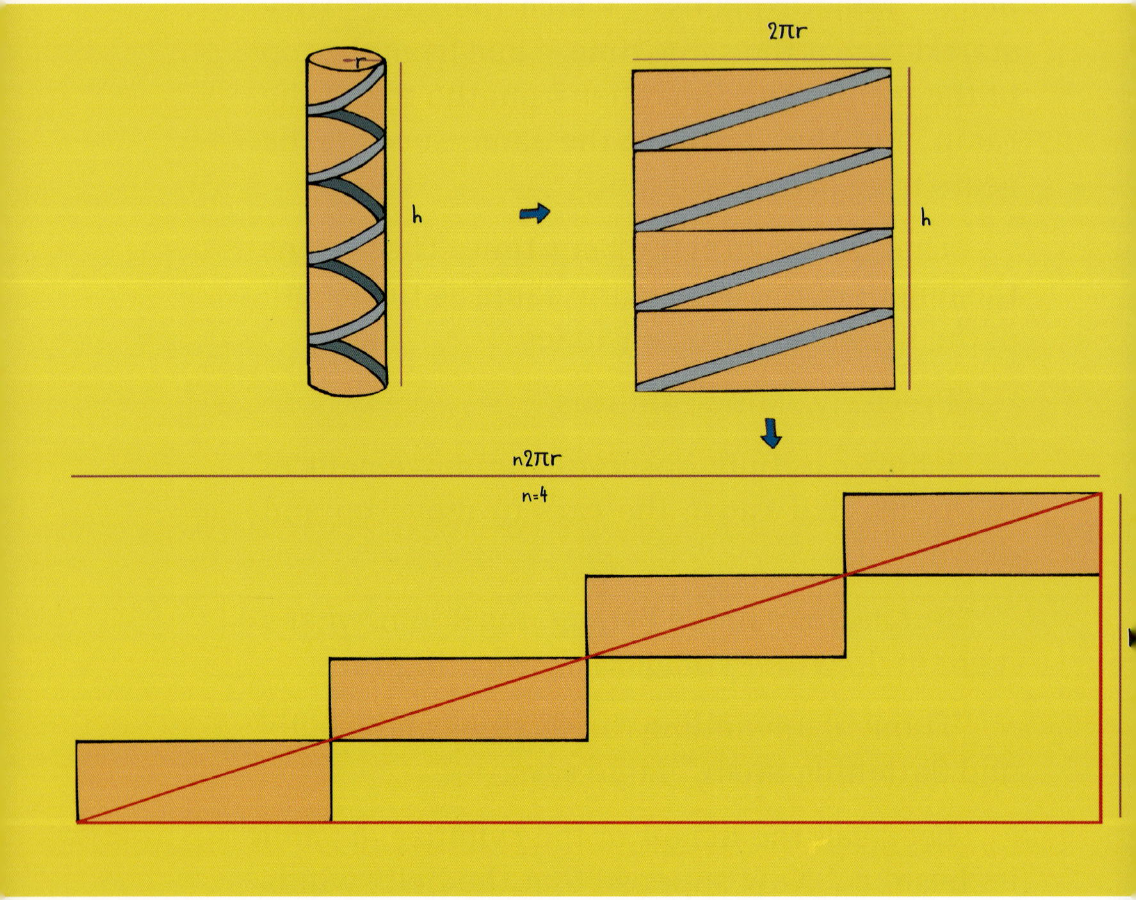

circumference of the base of the cylinder."

"If we assemble the flattened sections diagonally, the resulting configuration suggests a right triangle of height h and base $n2\pi r$."

"I get it," Jai says. "Then by the Pythagorean

theorem, the length of the incline is

"That's right!" Miho exclaims, admiring the

$$\sqrt{(n2\pi r)^2 + h^2}."$$

speed with which Jai picks up her explanations.

His friends are not surprised. Jai always understood things faster and better than them.

Math in Music

While inside the Pythagoras Room, they hear music coming from somewhere. Then they see a sign:

THIS WAY TO THE MUSIC ROOM

It is pointing to an unusual looking spiral staircase. Its steps are of unequal length and some steps are carpeted. A guide named Yasu tells them to go down the stairs one at a time. As usual Kino takes the lead. Each step he takes creates a musical tone. A melody emerges as he continues down.

Ichiro and Jai follow but they are curious to see what will happen if Ichiro goes ahead and Jai follows a few steps behind. They ask Yasu if they can do this and he consents. As they traipse down, a blending of tones is heard. The musical stairs are two stories high and actually bring them to the basement of the building.

At the base of the stairs are two posters, one about the staircase, another on music and mathematics.

Musical Staircase

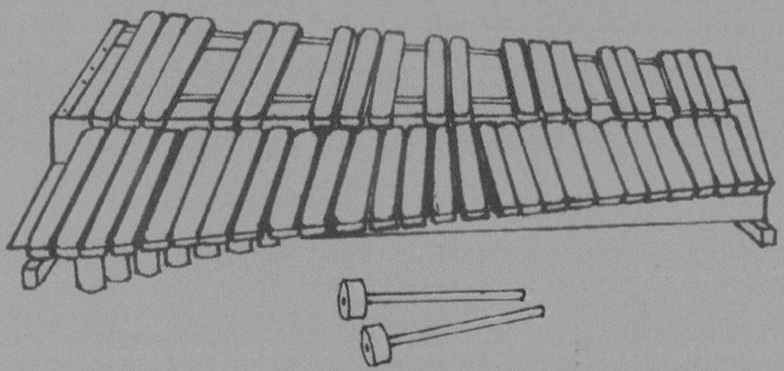

A xylophone is a musical instrument made of hardwood bars in graduated lengths set on a frame. The bars are each tuned to a different tone by length. The xylophone is played using two mallets.

Our musical staircase was crafted using the xylophone principle. The steps are made of wooden bars of different lengths to sound various notes of a chromatic scale. When you walk down the stairs you hear a simple melody.

The relationship between length and tone was known to Pythagoras. He observed the tones created using various lengths of string under different tensions. He created a one-stringed instrument, a monochord. Such instruments produce different tones when the string is pressed at certain lengths and plucked.

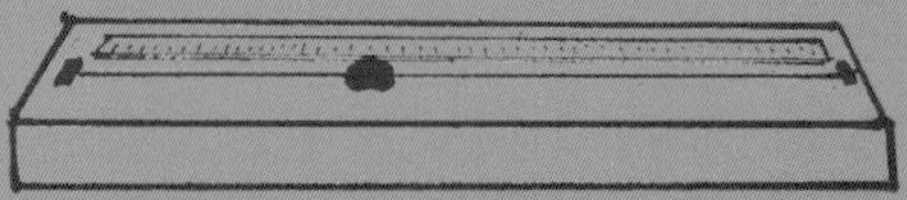

"Music is the pleasure the human soul experiences from counting without being aware that it is counting."

Gottfried Leibniz
(1646-1716)

Leibniz and Isaac Newton are both credited with the invention of the Calculus. Leibniz made significant contributions to fields other than mathematics. He was also a diplomat, historian, and philosopher among others. He is often described as "the last universal scholar."

On one side of the room is a piano. A guide, Chie, is playing the notes of the scale and explaining the relationships among the notes.

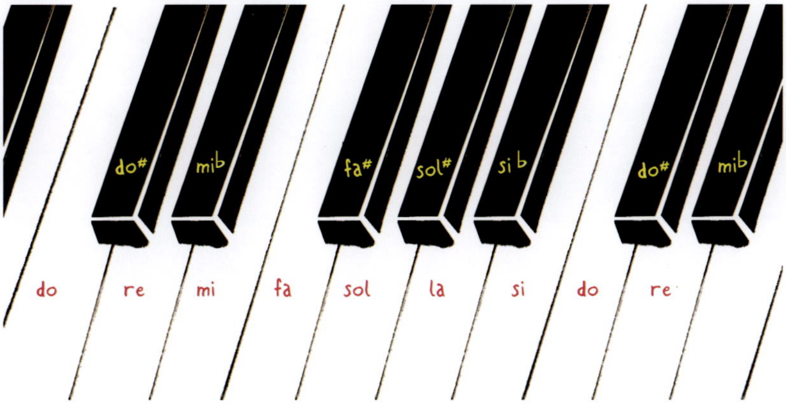

"I will play some chords and you tell me whether they are pleasing to listen to or not," she says. "Clap your hands if they are."

She plays do-mi-sol. They all clap.

She goes through a succession of chords. They clap again when she plays the chords do-fa-la and si-re-sol, going from the first to the second octave.

do
do#
re
mib
mi
fa
fa#
sol
sol#
la
sib
si

"Let's put the notes on a circular scale," she says leading them to a diagram. "Now let's look at those harmonious chords again and let's measure the distances between the notes in them."

"4 - 3 - 5 for do-mi-sol," Kino volunteers.

"3 - 4 - 5 for re-fa-la," another kid says.

"3 - 5 - 4 for si-re-sol," says yet another.

"Do those three numbers remind you of something?"

Jai is thinking hard, then blurts out "Those are lengths of sides of right triangles."

"Right!" Chie says. She brings out right triangles with side lengths 4 - 3 - 5, 5 - 4 - 3 and 3 - 5 - 4.

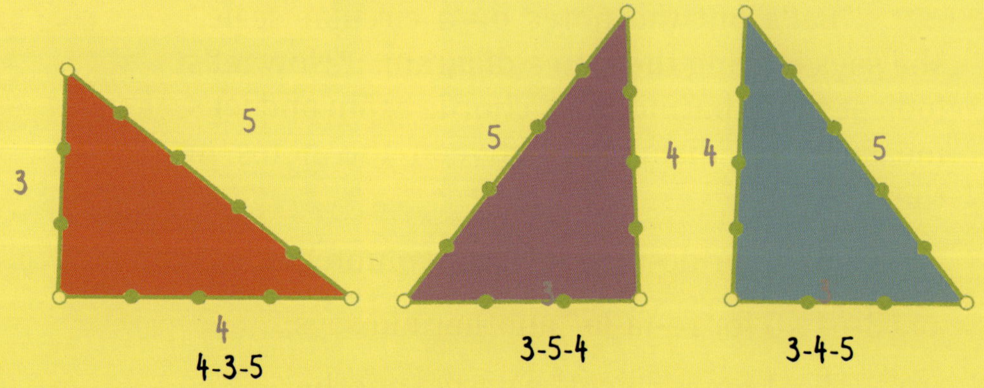

"As you can see the 4 - 3 - 5 and 3 - 5 - 4 triangles are congruent."

"So the harmonious chords are related to the Pythagorean Theorem?" Kino wonders out loud, amazed that such a connection exists.

"They sure are," Chie answers.

"But what if we play a chord with notes which are 4 - 3 - 5 apart but starting with do#? Will we still get a harmonious sound?" Ichiro asks.

"Try it out!" Chie answers.

Kino plays the chord do# - fa - sol# on the

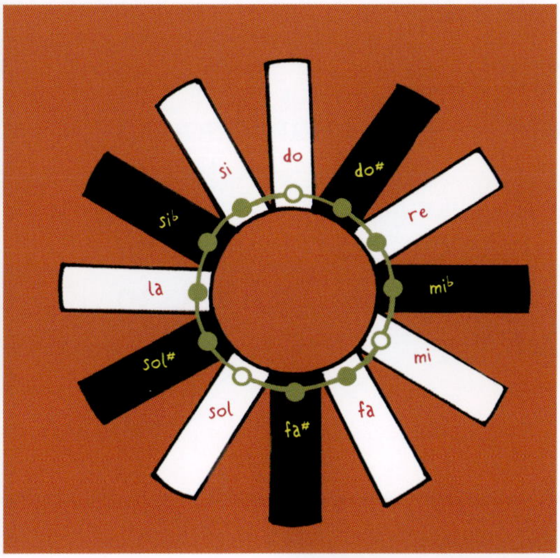
do
do#
re
mib
mi
fa
fa#
sol
sol#
la
sib
si

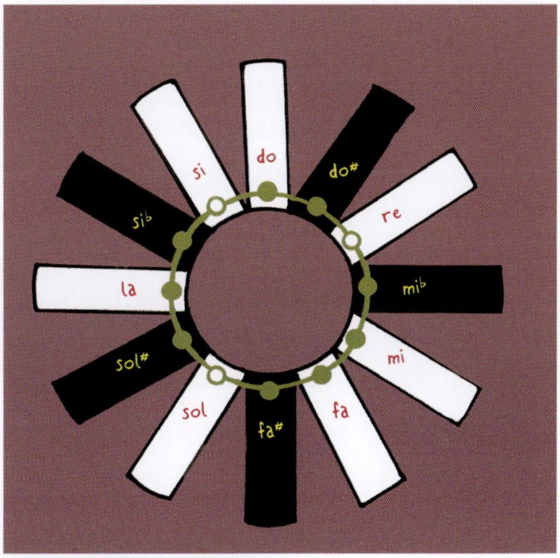
do
do#
re
mib
mi
fa
fa#
sol
sol#
la
sib
si

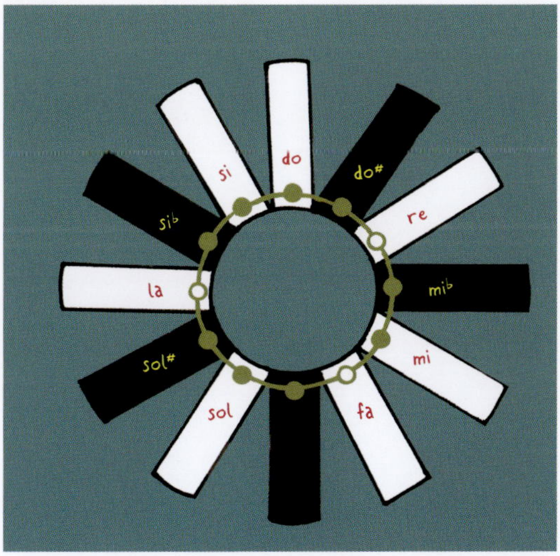
do
do#
re
mib
mi
fa
sol
sol#
la
sib
si

piano. The kids clap. They line up to take their turns to play the other chords.

"One more thing," Chie adds, "4 - 3 - 5 chords sound very upbeat while 3 - 4 - 5 chords sound melancholy."

The boys go to a huge circular xylophone striking various 4 - 3 - 5 and 3 - 4 - 5 chords to check out what Chie said.

Pachinko[2] Math

2. A gaming device which is a cross between a pinball and a slot machine, a player controls the flow of balls through an array of pins

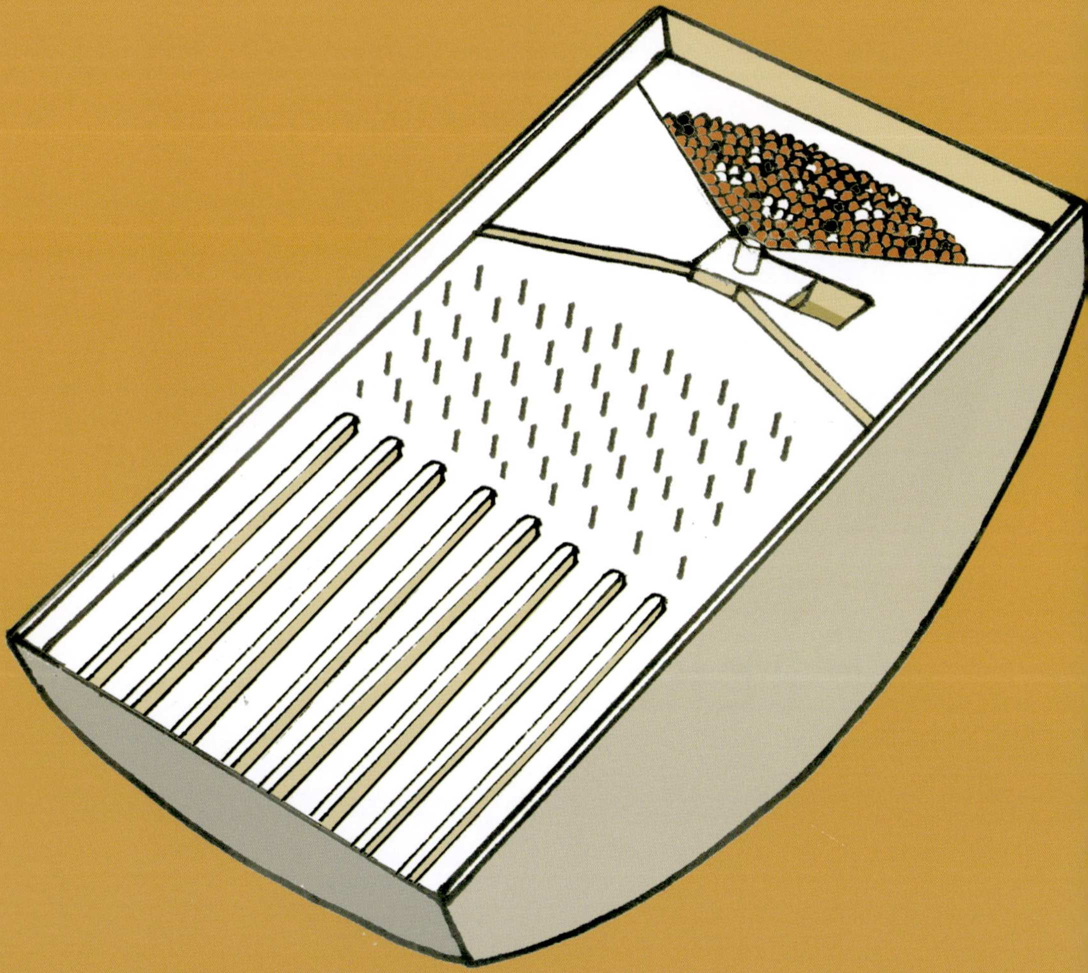

Just outside the Music Room is an open exhibition hall with many curious-looking objects.

On the floor is a rather large wooden object. Its base is like that of a rocking cradle. It is topped by a board with pins sticking out in an orderly fashion. Above the pins is a compartment holding plastic balls, and below the pins are some vertical compartments.

The boys have no idea what this is. They get on their knees to inspect it more closely. The pins are located at equal intervals horizontally. Each horizontal line is skewed at half intervals from the previous.

"It is like pachinko," Kino remarks.

One of the guides, Norio approaches them. "Tilt the device so that the compartment with the balls is higher," he instructs.

Ichiro and Jai tilt the device.

"Open the gate of the compartment and let the balls roll down," he continues.

Kino does this. They watch the balls roll down.

An arrangement that looks like a mountain appears along the vertical compartments.

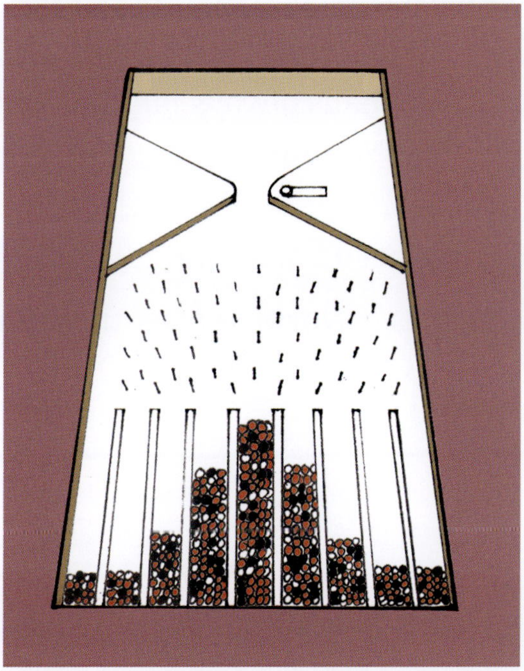

Intrigued, they repeat the exercise. First Kino tilts the device in the other direction so the balls return to their compartment.

"Same kind of arrangement," Ichiro observes.

"That's called a normal distribution," Norio informs them.

They do it three more times with the same

result. The last three times, Jai observed the balls very closely: A ball that hits a pin has equal chances of falling to its right or left.

"Why do the balls always fall that way?" both Kino and Ichiro want to know.

"Does it have something to do with the way the pins are placed?" asks Jai.

"Yes, it does." Norio senses their eagerness to learn. He leads them to a board and tries to explain with a diagram.

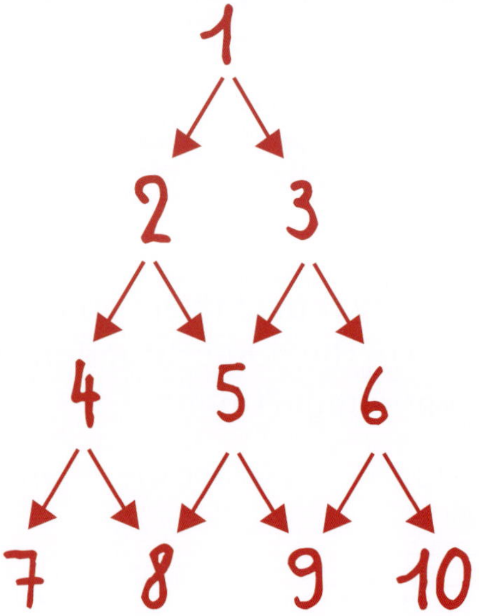

"A ball from the gate goes directly to the middle pin. When it hits that pin, it can fall to its right or left with equal probability. We can say the probability is in the proportion 1 : 1. So the probability of falling to the left, between, or to the right of the two lower pins is in the proportion 1 : 2 : 1."

"Another way of looking at it is to count the paths. There is one path to pin 2 and another one to pin 3, then looking at the next lower level, there is one path to pin 4, two paths to pin 5 and one path to pin 6. If we continue, there is one path to pin 7, three paths to pin 8, three paths to pin 9 and one path to pin 10. If we go on to the lower pins, the numbers build up around the center, so there are more chances that the balls will end up in the central compartments below."

He continues to write on the board.

"Now if I write the number of paths this way, what do you see?" he asks the boys.

1

1 1

1 2 1

1 3 3 1

1 4 6 4 1

"Those numbers written in a triangle like that look very familiar. Guys, where have we seen that before?" Kino asks trying to recall.

"It's Pascal's Triangle," Ichiro answers.

"You'll be surprised that so many things in mathematics are connected. Pascal's Triangle pops up again and again," Norio says.

GCF-LCM Machine

2
2
2
3
2
3
3
5

"Look at this," Ichiro calls out. "This machine will compute GCF and LCM automatically."

They had just learned greatest common factor and least common multiple in math class so they are quite interested in this machine.

"But only for those numbers whose prime factors are 2, 3 and 5, it says here," Ichiro adds, pointing to a note on the table.

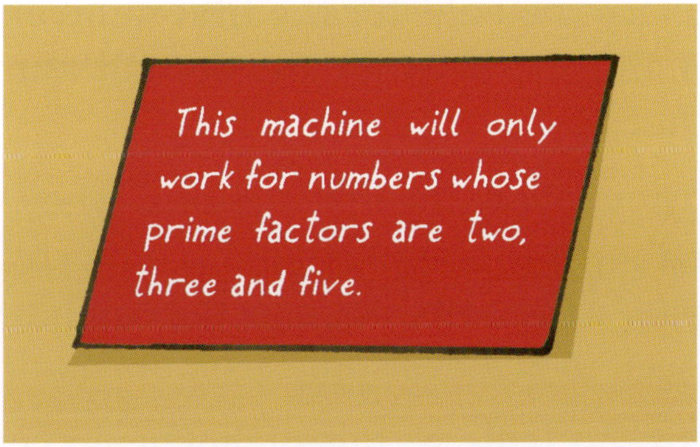

"OK," Kino says. "What do we have to do?"

They read the instructions.

1. Choose two numbers and factor them into primes.

They choose

$$a = 90 = 2 \times 3 \times 3 \times 5$$

and

$$b = 24 = 2 \times 2 \times 2 \times 3.$$

2. On the table are balls marked 2, 3 and 5. For each number, choose the balls that represent its factorization.

3. Place the balls for the first number in one funnel and the balls for the second number in the other.

4. They follow the instructions. They see the balls falling into compartments connected to the funnels. As the balls settle in, the compartments move as if to achieve some kind of balance.

Take the balls in the higher level compartments and multiply the numbers appearing on them to get the GCF. Do the same for the balls appearing in the lower compartments to get the LCM.

So they find

$$GCF = 2 \times 3 = 6$$

$$LCM = 2 \times 2 \times 2 \times 3 \times 3 \times 5 = 360.$$

They know these are the right results; but Ichiro just has to know how this machine works. He approaches a guide, who introduces himself as Minoru.

Minoru is glad to explain. "When you place the balls in the funnels, they are first sorted by size. Below each funnel is a board with three holes whose sizes match the sizes of the balls. These holes are arranged in increasing size so as the balls pass through, only balls of weight 2 will fall through the first hole, only balls of weight 3 will fall through the second hole and so on. The balls of the same size from both numbers are sent into adjacent compartments. There is a see-saw mechanism which will make the compartment with fewer balls rise and the one with more balls fall."

"So the higher compartments will contain the factors common of the two numbers and when you multiply them, you get the GCF." Ichiro concludes, pleased with his grasp of the explanation. "And the lower compartments have the larger number of each kind of factor appearing in either number and so their product is the LCM," he continues.

"So the machine does the work for us," Kino says. "We should have one of those machines in math class!"

Jai probes further "What if we choose two numbers with an equal number of factors of the same kind?"

"Yes, " Ichiro picks up the question, "Suppose we choose

$$a = 90 = 2 \times 3 \times 3 \times 5$$

and

$$b = 60 = 2 \times 2 \times 3 \times 5."$$

"Well those equal number of factors will appear in both the GCF and the LCM," Minoru responds. "So a small weight has been placed so that the right compartment will appear lower, but this weight is very small and will not affect the general case." They try it out.

Sure enough, the right compartment with one 5 appears slightly lower than the left compartment with one 5.

"So $GCF = 2 \times 3 \times 5 = 30$

$LCM = 2 \times 2 \times 3 \times 3 \times 5 = 180$."

Ichiro is impressed with this machine. Looking at all the balls in the compartments, he says, "Now I understand that formula

$$a \times b = GCF \times LCM$$

we learned in class."

"How does that work?" Kino asks.

"See, all the balls we put in the funnels are either in the GCF or LCM; when you multiply GCF and LCM together, you get the product of the two numbers," Ichiro explains.

Minoru is pleased to note that the machine has helped the boys understand their math lessons. They thank him and look around for their next destination.

Baumkuchen,[3] Spaghetti & Watermelon

3. A traditional German cake made from many layers of thin rings.

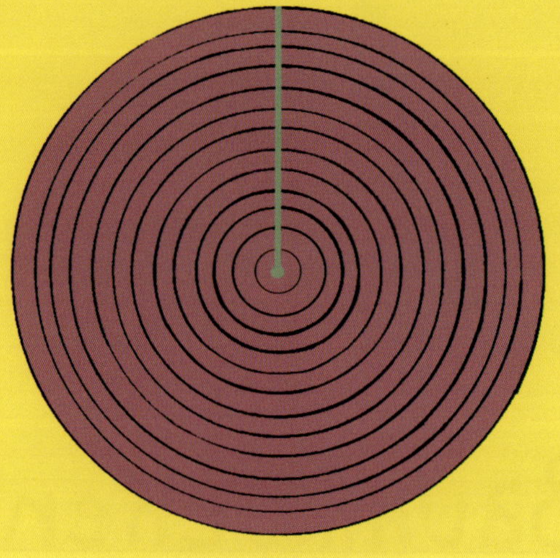

There are more exhibits in the hall. One table is marked

AREA OF A CIRCLE

and on it is an object that looks like a round cake made up of circular layers.

"*Baumkuchen*," says Kino.

Actually the circular model is made up of many layers of velcro.

A guide named Kyoko approaches them and asks, "Do you know how to find the area of a circle?"

"No," they answer.

"We didn't get to that part yet in our math class," Kino explains.

"How about the area of a triangle?"

"We just learned that," Ichiro volunteers,

$$\text{"}\frac{1}{2} \times \text{base} \times \text{height}\text{."}$$

"Right," Kyoko says. "Well, it is always a good strategy to reduce something you don't know to something you know."

"We will just transform the circle into a triangle," she says. She cuts the *baumkuchen* along a radius and produces what looks like a triangle.

"So now, the area of the circle is the area of this triangle."

"What is the base of this triangle?" she asks.

"It's the circumference of the circle, but we haven't taken that formula up either," Jai says.

"Let me tell you then," Kyoko offers. "The circumference of a circle is $2\pi r$ where r is its radius and π is a very special number whose value is approximately 3.14."

"The height of the triangle is the radius of the circle, so if its base is $2\pi r$, that makes its area

$$\frac{1}{2} \times 2\pi r \times r = \pi r^2,"$$

says Ichiro.

"Good! I hope you will remember that formula since you figured it out by yourselves."

Jai is not quite convinced.

"The figure looks like a triangle but the sides are somewhat jagged, won't that make a difference?" he asks.

"It will help to think of the layers as being paper-thin," Kyoko answers as she leads them to a board and draws a diagram to explain.

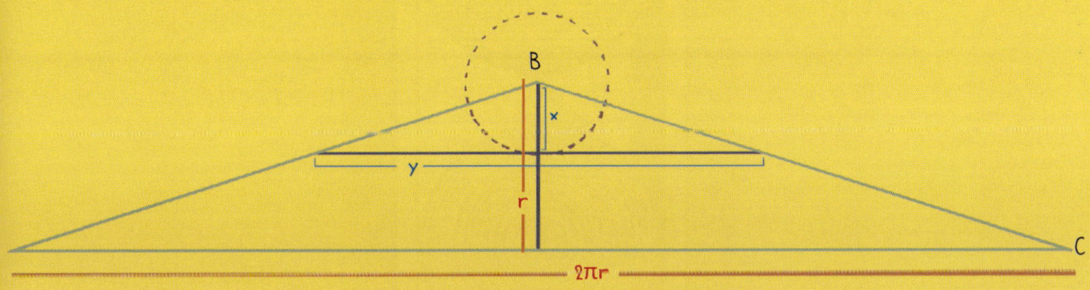

"The length of the base of the triangle is $2\pi r$ as we already said, where r is the radius of the original circle. Let us consider a portion of the original circle. Suppose its radius is x less than r. Then y will be the circumference of that smaller circle and so what this tells us is that the y values are proportional to the x values. Since the x values change continuously throughout the figure, so do the y's. So there are no jagged edges in AB and BC," she continues.

Jai nods to show he understands.

On another table is a model consisting of a hemisphere whose surface is covered with a length of plastic tube. The tube extends to line the interiors of two circles directly below. The circles have the same radius as the hemisphere.

The label for this table reads

SURFACE OF A SPHERE

"What we have here is a bowl and two plates of spaghetti," Kino jokes.

Kyoko pumps blue liquid into the tube until the surface of the hemisphere is completely covered. Then she makes the liquid move down to the circles. The surface of the hemisphere is completely empty when the areas of the circles are completely filled.

Jai notes that this is the same principle as one of the Pythagorean models. "Let me guess," Jai volunteers.

"Surface area of a hemisphere = 2 × area of the circle."

"Yes, as long as they have the same radius," Kyoko confirms.

"Therefore,

$$\text{surface area of a sphere} = 4 \times \text{area of the circle}$$
$$= 4\pi r^2\text{,"}$$

Jai continues.

The next table is labeled

VOLUME OF A SPHERE

and on it is a wooden model painted to look like a watermelon.

Ichiro discovers that the watermelon can be broken up into cone shaped slices.

"It's sliced for eating," Kino observes.

Ichiro and Jai look at the model thoughtfully.

"So the volumes of the slices add up to the volume of the watermelon," Ichiro says.

"You are getting the hang of it," Kyoko says encouragingly. "The entire watermelon is a sphere and the slices are cones whose heights equal the radius of the sphere."

"But you'll have to help us with this again because we don't know the formula for the volume of a cone," Jai requests.

"The

$$\text{volume of a cone} = \frac{1}{3} \times \text{height} \times \text{base area,"}$$

she says.

"In this case the height of each cone is *r*. So the

$$\text{volume of a sphere} = \frac{1}{3} \times r \times (\text{sum of the base areas of all cones})."$$

"But the base area of all cones is the surface of the sphere," Ichiro says.

"And we already know the formula for that from the previous model," Jai recalls. "So

$$\text{volume of a sphere} = \frac{1}{3} \times r \times 4\pi r^2$$

$$= \frac{4\pi r^3}{3}."$$

Ichiro and Jai are pleased with themselves for being able to arrive at the formulas. Although he did not contribute to the discussion, Kino was listening intently and he understood.

"*Baumkuchen*, spaghetti, watermelon -- all this talk of food makes me hungry. Let's go for lunch," Kino suggests.

Automat(h)

The boys are very hungry but they don't want to spend too much time for lunch.

"Let's eat a quick lunch so that we have more time for the exhibits," Ichiro suggests.

"Is there a place where we can eat?" they ask a passing guide.

"Over there," he points to a place called

AUTOMAT(H)

They enter and all they see are vending machines. It slowly dawns on them why the place is called Automat(h).

An attendant tells them they have to buy tokens for the vending machines. They are amused to see the tokens.

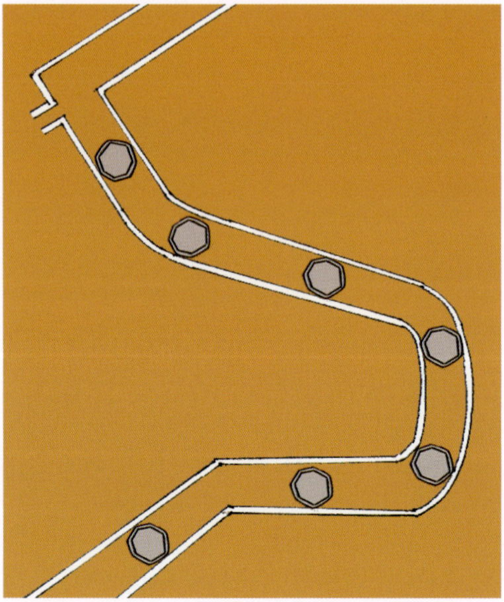

“Shapes of constant width!” Ichiro notes.

When they get to the vending machines and put in a token, they can actually see it move down a pipeline.

“Why do you think they chose to make tokens like these?” Kino wonders.

“To show they are just as good as round coins for vending machines,” Jai offers an explanation.

They see *onigiri*[4] in one of the vending machines.

"Fat triangle!" they exclaim.

When they have bought their sandwiches and drinks, they go to the eating area. As soon as they sit down, they see posters on the wall facing them, each with a math problem about food.

While they eat, they begin to discuss the first problem, drawing diagrams on paper napkins.

4. Traditional Japanese rice ball shaped like a Reuleaux triangle.

If you were to choose between two smaller square pizzas S and M or a bigger square pizza L, which would you choose?

"Those squares remind me of the Pythagorean Theorem," Kino remarks.

"You're right, what if we arrange them around a right triangle to see if they fit together?" Ichiro says.

This is how their solution turns out.

If

then area (S+M) < area L, so choose L.

If

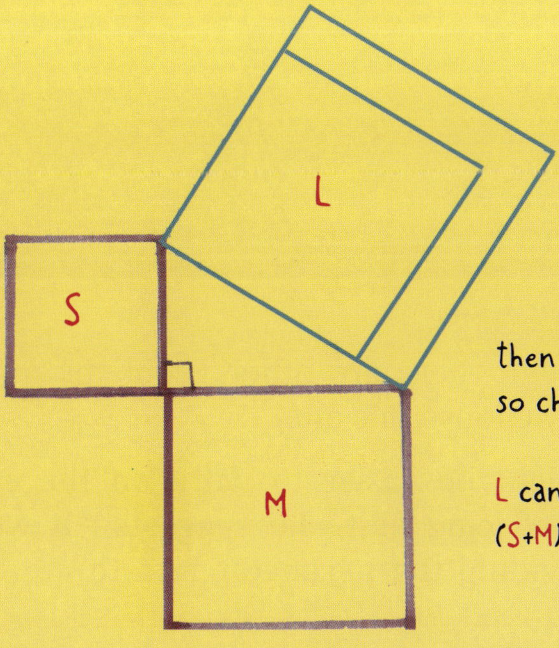

then area (S+M) ≥ area L, so choose S and M.

L can also be chosen if area (S+M) = area (L).

A rectangular cake has the given dimensions.

It has icing on top and on all four sides. Divide it into three pieces of equal volume in which the horizontal cross-section of each piece is a quadrilateral, and so that each piece has the same amount of icing as the others.

Although there are other problems on the wall, they choose this one.

"It might be easier to work on the equal amount of icing first," Jai suggests, "It would mean dividing the perimeter into three equal parts."

They compute:

Starting from one corner, they divide the perimeter

$$\text{perimeter} = (2 \times 45) + (2 \times 30) = 150$$

$$150 \div 3 = 50.$$

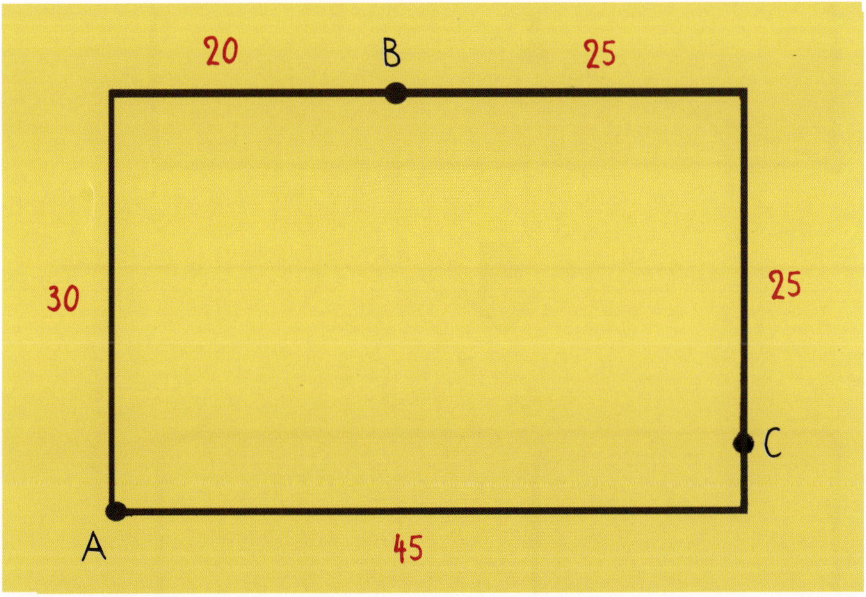

"The area of the rectangle is

$$30 \times 45 = 1350,$$

so if we just look at the top of the cake, the area

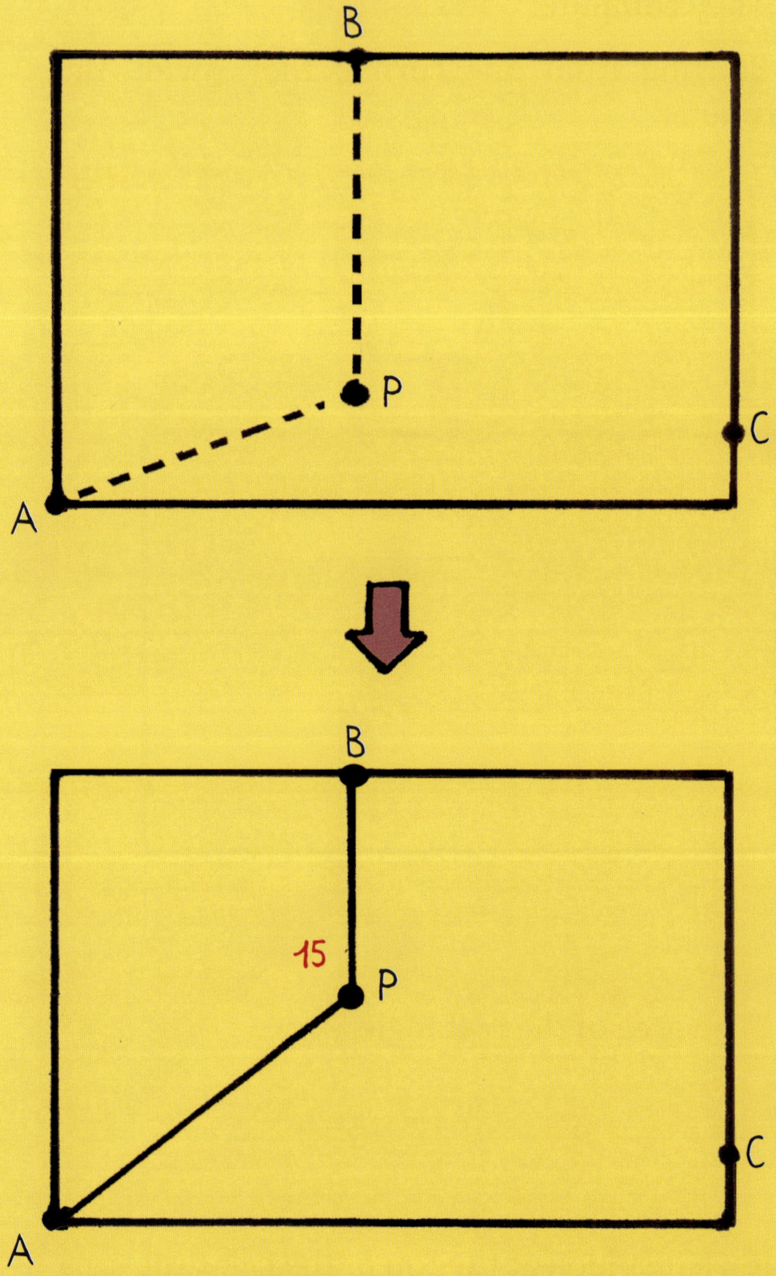
B
P
C
A
B
15
P
C
A

of the top part of each piece must be

$$1350 \div 3 = 450,"$$

Ichiro says.

"We also know that to get equal icing, one piece must contain the perimeter from B to C, another piece must contain C to A and the last piece A to B," Kino says.

"How do we find a quadrilateral including the perimeter from A to B?" Ichiro asks.

"The two other sides must be inside the rectangle," Jai replies. He is thinking very hard. "But for what kind of quadrilaterals can we find the areas right away?" he asks.

"A trapezoid might do," says Ichiro.

"Let's try it." says Jai.

He adds to the diagram and says "We should find the point P so that the area of the quadrilateral is 450."

"Well, let's use the formula for the area of a trapezoid," Ichiro suggests.

$$\text{"}\frac{1}{2}(b_1 + b_2)h = 450,$$

or

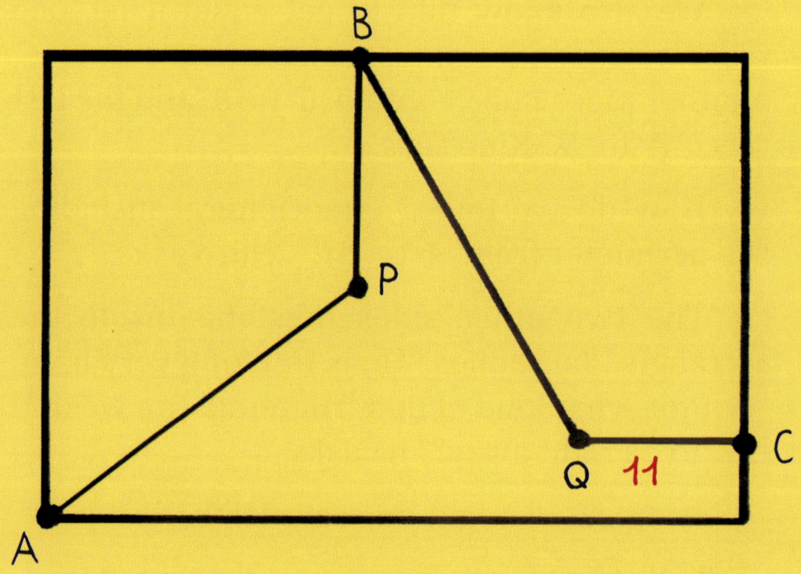
B
P
Q
11
C
A

$$\frac{1}{2}(30 + b_2)20 = 450."$$

They solve for b_2 and find $b_2 = 15$. Ichiro adjusts the diagram to show P in the right position.

"We can do the same to find a trapezoid whose perimeter includes the piece from B to C," Jai suggests.

They use

$$\frac{1}{2}(25 + b_2)25 = 450$$

to find $b_2 = 11$. They add to their diagram.

"We have three pieces now with the same amount of cake and icing but the third piece is not in the shape of a quadrilateral," Ichiro says discouraged.

They don't know how to proceed and are ready to give up, but then Kino sees one of the guides at the vending machines.

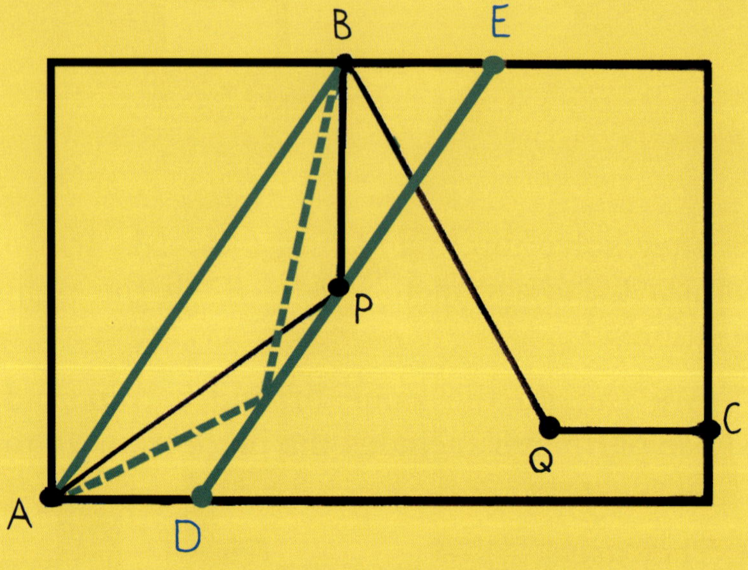

"Maybe we can ask the guide to help us," he says.

Kino approaches the guide. He comes back with Kino to their table and they explain their problem.

"I'm Yasu," he says. "This is not an easy problem," he continues in a sympathetic tone. "What you need to do is draw a line from A to B and look at the triangles with base AB and the same height as $\triangle APB$. Any of these triangles would have the same area as $\triangle APB$."

He draws the line AB and a line DE parallel to AB through P on their diagram.

"You can actually move the point P anywhere along DE and get a triangle whose area is equal to $\triangle APB$, so you could replace $\triangle APB$ with another triangle," he continues.

"Now, do the same thing with the other trapezoid," he instructs.

Ichiro draws the line BC and a line FG parallel to it through Q.

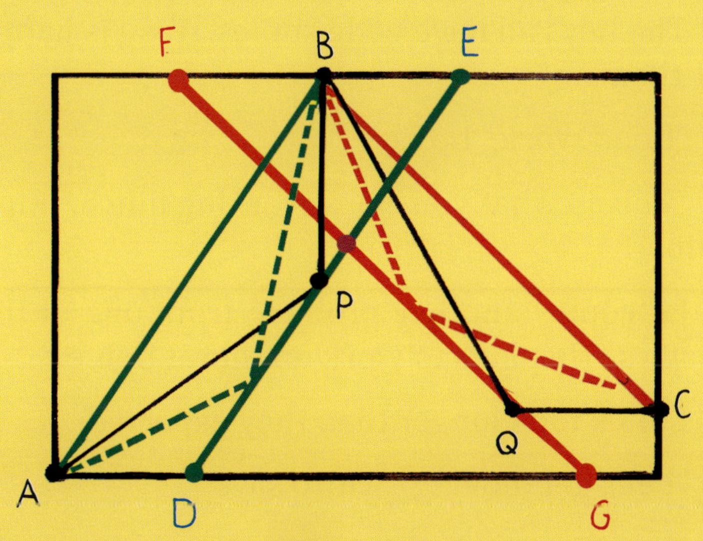

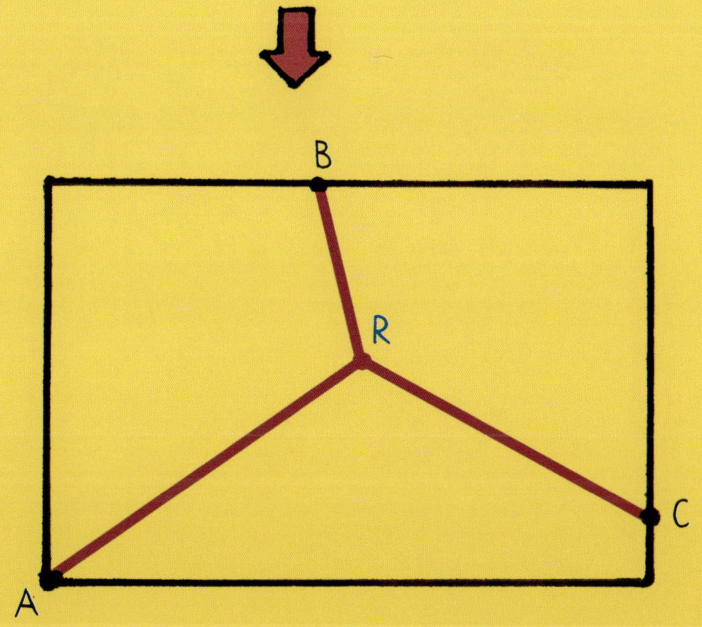

"So you can move the vertex P along DE and the vertex Q along FG," Jai observes.

"I get it!" Ichiro exclaims. "We can move both P and Q to the intersection R of DE and FG. That way the third piece will also be a quadrilateral."

The boys all have wide smiles. Yasu is happy for them.

"Good work," he says.

"This is what you call a working lunch," Kino quips.

Jai copies another problem from one of the posters. He plans to work on this at home.

Lunch took longer than they expected.

"Let's get moving," Ichiro says.

Calisson is a French confectionery made of almond and melon paste. Each piece looks like two equilateral triangles joined along an edge.

Suppose calissons of side length 1 are packed into hexagonal boxes of side length 3 as shown.

There are only three orientations possible when laying a calisson in the box,

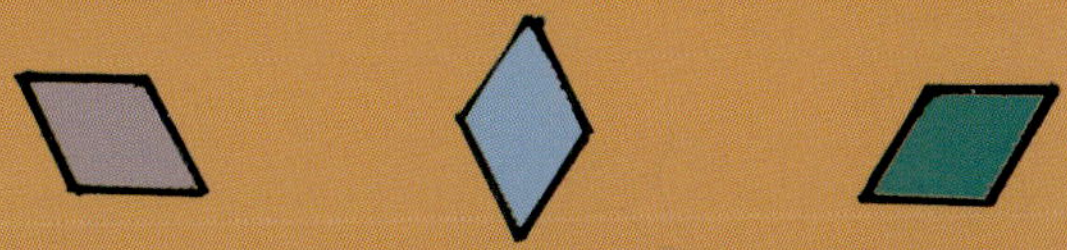

but many ways of packing the box.

Find several different ways of packing the box with calissons. For each packing, count and compare the number of calissons of each orientation. What do you observe?

A Slice of a Cone

Close to the cafeteria is a room whose entrance attracts their attention. Four huge plastic double cones guard it. As they approach they see that each cone is cut by a plane and the curves formed at their intersections are clearly visible. A sign on a wall facing them reads

CONIC SECTIONS

Inside the room is a poster explaining what conic sections are.

Conic Sections

A conic section is a curve formed by the intersection of a plane with a double right circular cone. The inclination of the plane, in relation to the vertical axis of the cone, determines the resulting conic-either a circle, an ellipse, a parabola or a hyperbola.

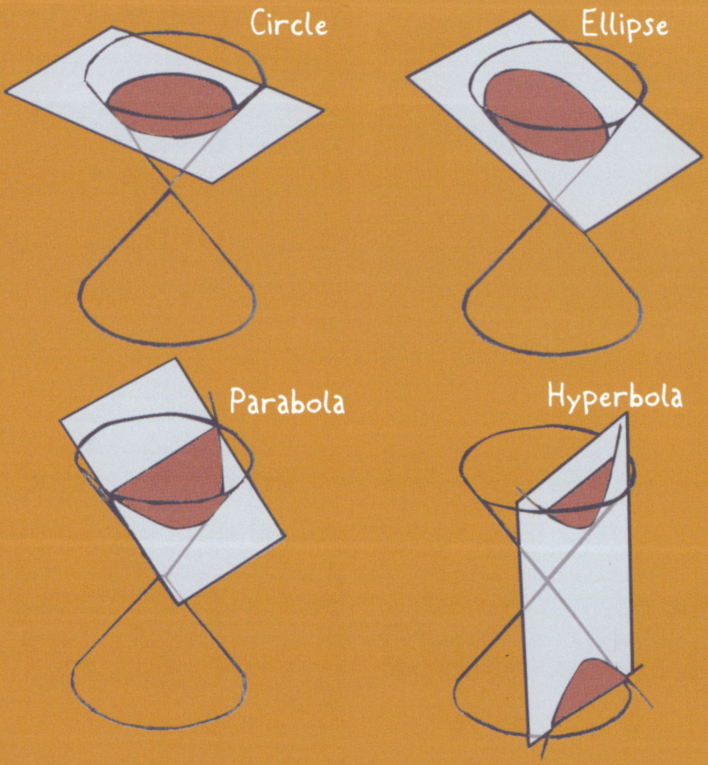

The study of conic sections dates back to the ancient Greek geometers Euclid (c. 300 B.C.), Archimedes (c. 287-212 B.C.), Apollonius (c. 260 - 190 B.C.).

Conic sections have played a vital role in mathematics and its applications since the time of the ancient Greeks. Johannes Kepler (1571-1630) discovered that the planets revolve around the sun in elliptic orbits rather than circular orbits, as had been believed. In 1672, Sieur Cassegrain invented a telescope that used the reflective properties of both hyperbolic mirrors and parabolic mirrors, a design incorporated in the Hubble telescope. The British astronomer Edmund Halley (1656-1742) used facts about elliptic orbits to predict that "Halley's" comet would appear every 76 years.

They are not familiar with most of the curves except the circle, but they are encouraged by the sight of many other kids who seem to be playing or experimenting with the rather large models in the room.

They come upon what looks like a billiard table. Its shape is what they now recognize as an ellipse.

"This is a billiard game you can never lose," a guide named Satsu approaches them.

"How do you play it?" Kino asks.

"An ellipse has two special points in it, each is called a 'focus'. Let's place a ball in each focus."

"Now you can make a bet with anyone that if you use the cue to hit one of the balls, you will also hit the other one," Satsu continues. "Try it out."

Ichiro takes a cue, hits one ball and sure enough, the ball rebounds from the boundary of the ellipse and hits the other ball.

Kino takes his turn and then Jai.

"Why does it work?" the boys ask.

"The ellipse has certain reflecting properties because of its shape," Satsu says as he leads them to a poster on the ellipse.

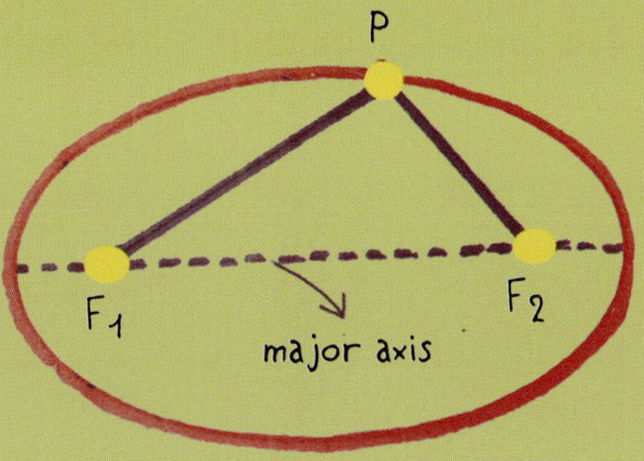

Ellipse

An ellipse consists of the set of all points P in a plane such that the sum of the distances of P from two fixed points in the plane, F_1 and F_2, is constant. Each of the fixed points is called a focus.

Reflective Property

A ray of light originating from one focus and directed at a point P on the ellipse will be reflected from P along a line segment that passes through the second focus.

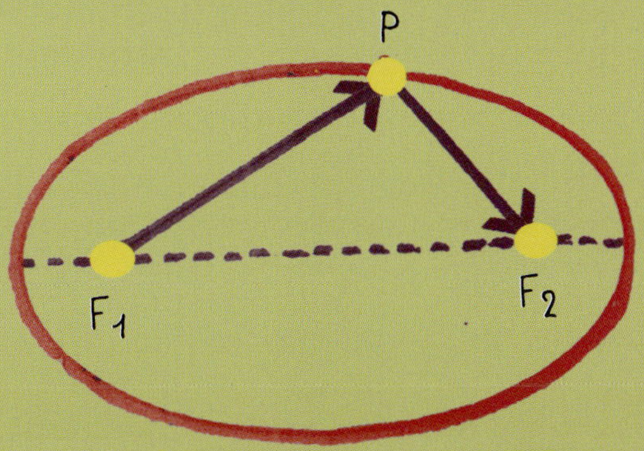

"Whenever you hit the ball at F_1, no matter which point it hits on the boundary of the ellipse, it will always be reflected back to F_2. Other conic sections also have special reflecting properties," Satsu states. "Let me show you a model that demonstrates the reflecting properties of a parabola."

He leads them to a table and they see what looks like a toy. It shoots balls in the direction of a curve, which they can now identify as a parabola, and the balls are always reflected to a hole into which they fall.

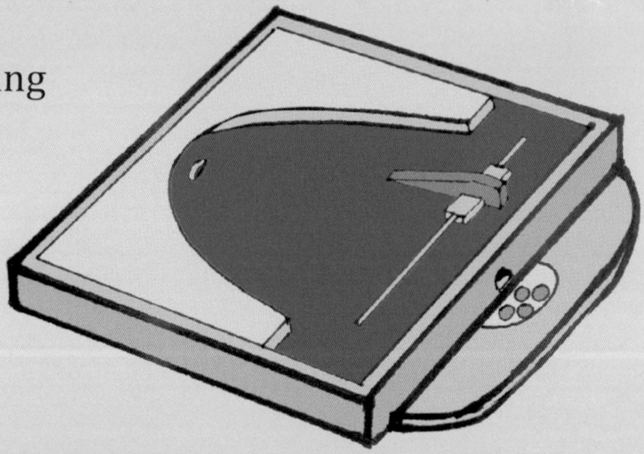

Jai is thinking hard. "In the billiards game, the ball is always reflected to a focus, is that hole also a focus?"

Satsu is taken aback at how clearly Jai has seen the analogy.

"Yes, " he replies, "A parabola has one focus and that hole is at the focus."

He shows them a poster on the parabola.

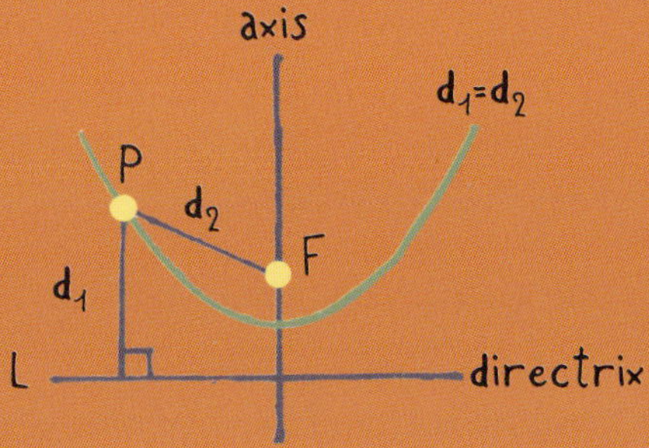

Parabola

A parabola consists of all points P in a plane which are equidistant from a fixed point F called the focus and fixed line L called the directrix.

Reflective Property

A ray of light parallel to the axis of symmetry of the parabola, directed towards the parabola, will be reflected off the parabola to the focus. If a light source is located at the focus, then light from it will reflect off the parabola in a straight beam.

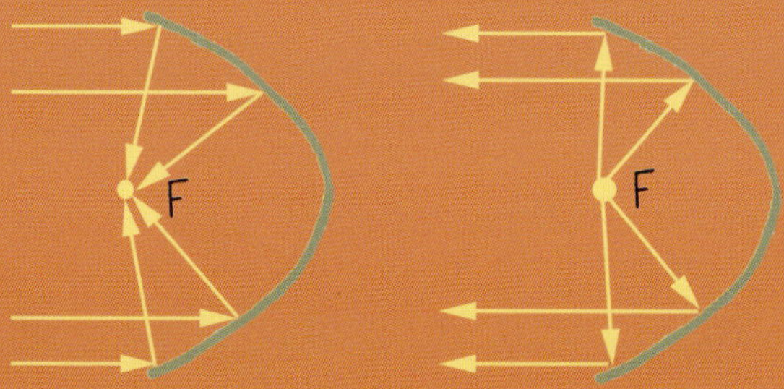

"Do you have a parabolic antenna for your TV set at home?"

"Yes," they all answer.

"Well, its shape is a paraboloid and it works in pretty much the same way. The antenna concentrates light and sound rays at the focus so your TV set can have better images and sound."

The boys are surprised by this statement, since they never imagined mathematics has anything to do with clear images on their TV screens.

"How do you get a paraboloid from a parabola?" Jai asks.

"Just rotate the parabola 180° around its axis." Satsu replies.

Satsu leads them to a collection of real appliances involving the reflecting property of the paraboloid. They see a parabolic antenna, a halogen heater and a car headlight.

Nearby two kids are dropping balls simultaneously from the same height into a paraboloid, and the balls always collide at approximately the same point. The boys watch this experiment for some time. They recognize this point of collision as the focus.

"We have an outdoor experiment involving the paraboloid, want to see it?" Satsu asks.

They climb a staircase leading to the grounds of Wonderland. Satsu leads them to an area where the paraboloid is located. The boys look around and see that the grounds are also organized as an outdoor exhibition space with a number of large objects displayed.

“There’s a lot we haven’t seen yet,” Kino remarks.

“We can go there after we see everything inside,” Ichiro says.

“If we have time today, there’s a lot we haven’t yet seen inside,” Jai says.

Satsu shows them a paraboloid with a potato at its focus. He adjusts the paraboloid so that the rays of the sun hit from a direction parallel to its axis.

“The potato will be roasted in a while,” Satsu says.

In the meantime, he shows them an actual outdoor cooker. It has an adjustable paraboloid base which can be moved to face the sun and a plate which can hold a kettle or a frying pan.

"Let's boil an egg," Satsu says filling a kettle with water and putting in an egg.

The boys are enjoying the experiments. Soon the skin of the potato changes color and the water boils.

"These are good for picnics," Kino says.

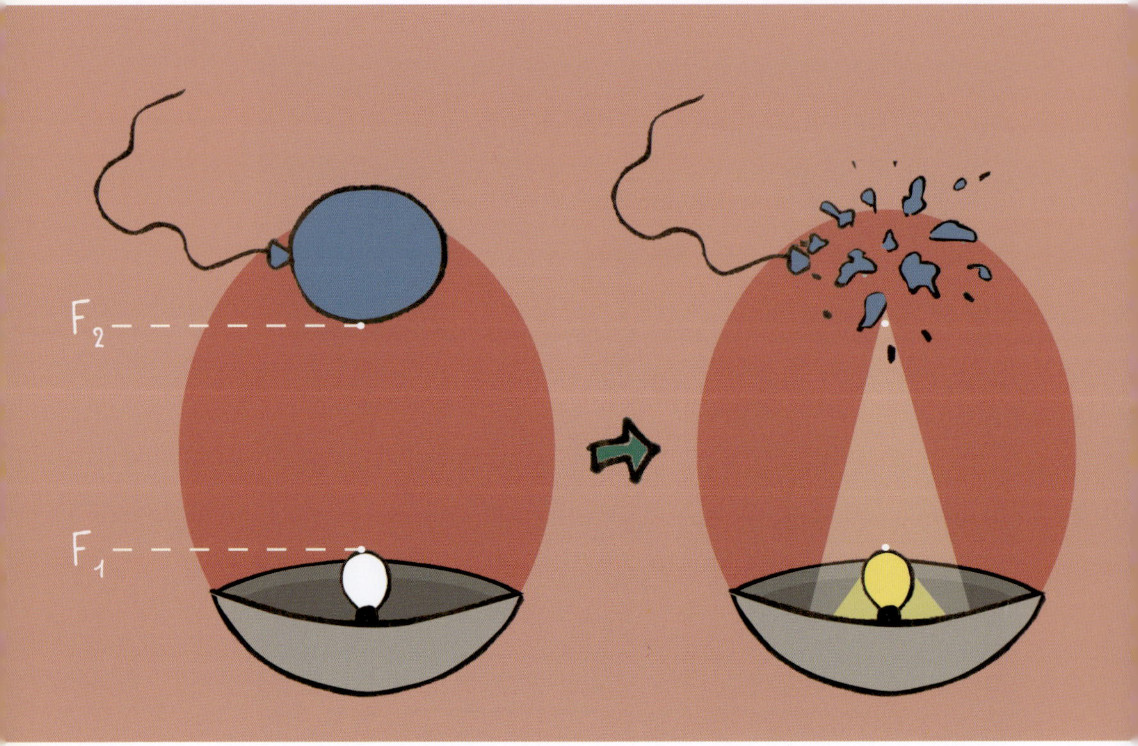

Satsu gives them the roasted potato and the boiled egg and leads them back to the conics exhibit.

“Come over here,” Satsu invites them.

They see a bowl with a light bulb at its center.

“This is a part of an ellipsoid,” Satsu says. “On one focus we have a bulb.”

"What does it do?" Ichiro interrupts.

"Why don't you inflate this balloon and place it where the other focus would be located?" Satsu suggests.

Ichiro does so. Satsu turns on the light and the balloon bursts at once.

"Can you explain what just happened?" Satsu asks them.

"I think the light from one focus was reflected by the part of the ellipsoid below it and concentrated on the other focus, where the balloon was, and the heat generated burst the balloon," Ichiro answers.

"Perfect," Satsu praises.

"They use this same principle in hospitals now to break up kidney stones," Satsu continues.

"How's that again?" the boys ask, not entirely understanding what he said.

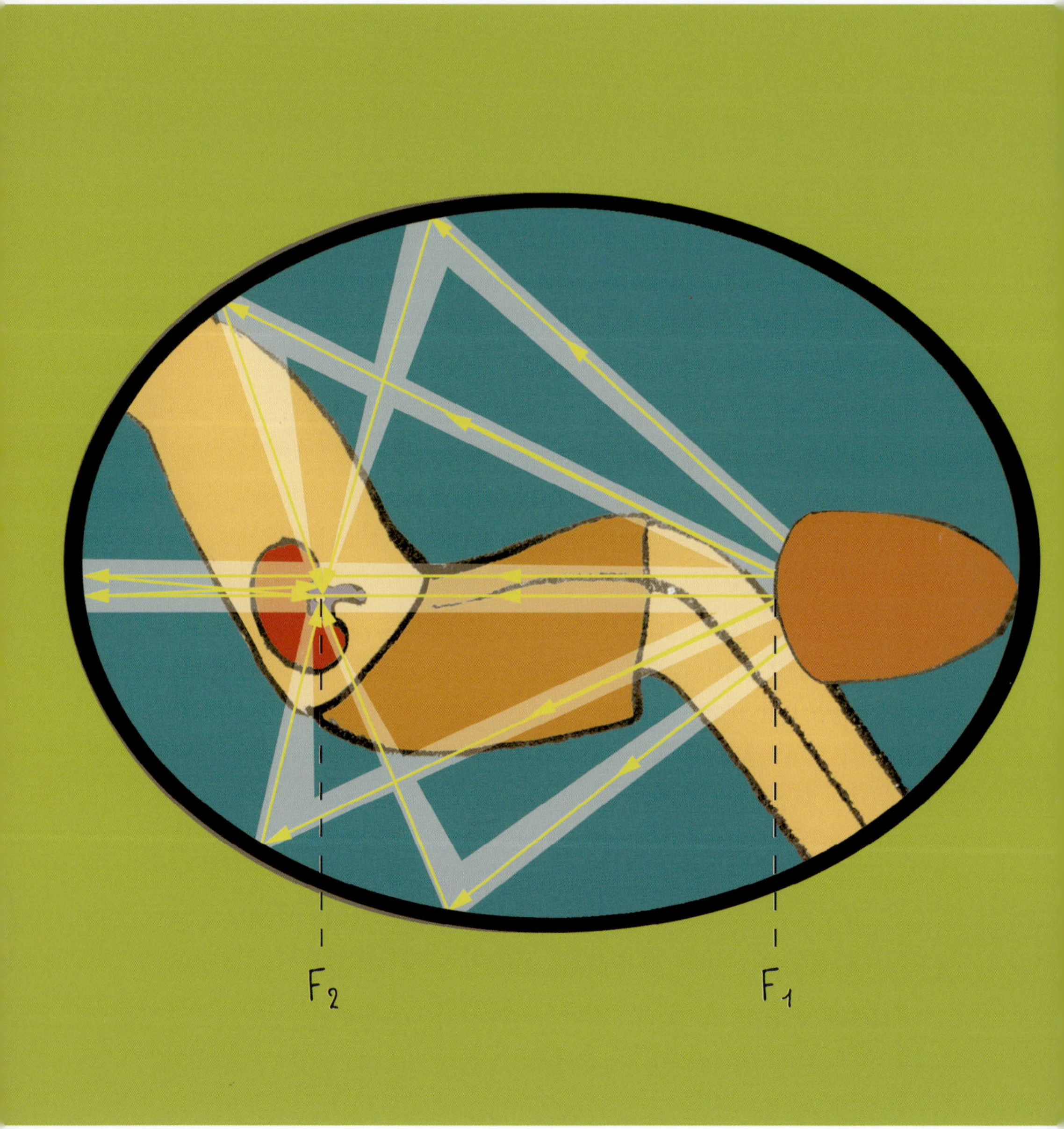

F_2
F_1

"Kidney stones can cause a lot of pain and infection too. Before, the only remedy was to operate, but now there is a fairly new medical procedure called ESWL -- that stands for exracorporeal shockwave lithotripsy."

"Wow! That is a mouthful," Kino comments. "Say it again, slowly please."

"Extracorporeal shockwave lithotripsy," Satsu repeats. "The lithotripter machine has an ellipsoidal part. The doctor positions the patient so that his kidney stones are on one focus of the ellipsoid. Then the doctor sends sound waves originating from the other focus. The sound waves bounce off the ellipsoid and are reflected to the position of the kidney stones. The sound waves shatter the stones, so there is no need for surgery."

Jai imagines what the machine would look like.

"So you have to know math too to be a doctor?" Kino asks.

"Doctors need a lot of math nowadays. Networks are used to track the growth of tumors, differential equations to model the spread of epidemics, and there's a machine called MRI that gives better images than x-rays nowadays, you have to know some calculus to interpret the pictures accurately," Satsu answers.

The boys are impressed although they have no idea what kind of math Satsu is talking about.

"What is that?" Ichiro asks, attracted to a rotating model.

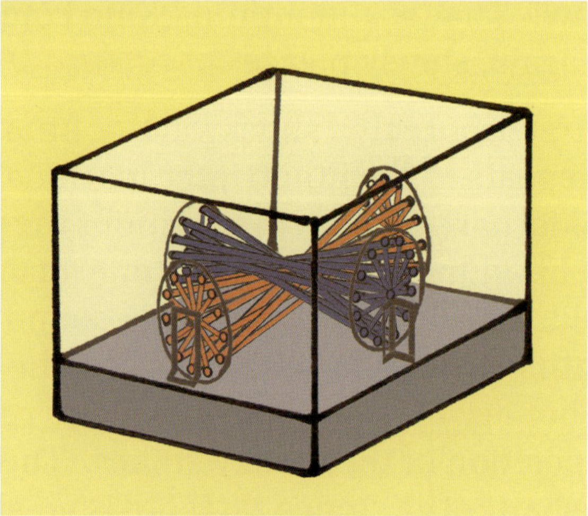

"This model illustrates how gears change the direction of motion from one axis to another. Certain types of gears, found in cars and other machines, are based on hyperboloids. Here we have two hyperboloids that are tangent to one another along a line. When one of them rotates around its axis, the other is forced to rotate around its axis, transforming motion in one direction to motion in another direction," Satsu explains.

Nearby is a poster on hyperbola.

$|d_1=d_2|=$ Constant

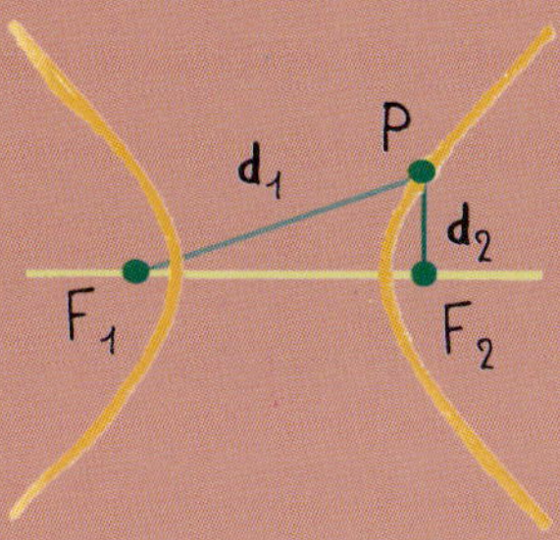

Hyperbola

A hyperbola consists of all points P in a plane such that the absolute value of the difference of the distances of P to two fixed points F_1 and F_2 in the plane is a constant.

Reflective Property

A ray of light from a focus will reflect of the hyperbola as if originating from the other focus.

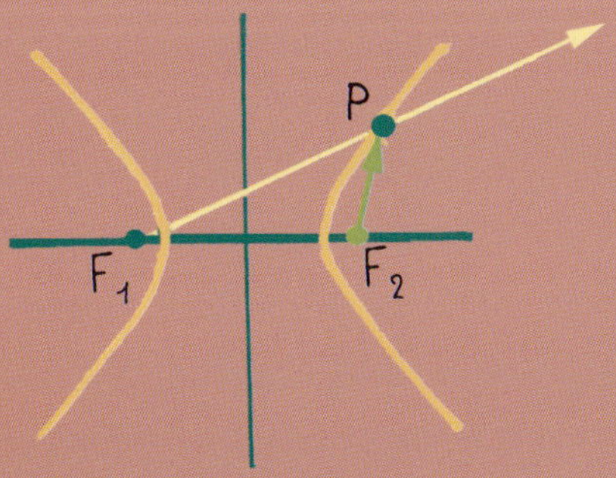

"Thanks, Satsu," they say, as they find their way out.

"Why don't we go and see the other models outside." Kino suggests.

"But we haven't even seen the stuff on the first floor yet," Ichiro protests.

"We can always come back another day," Jai says. "Why don't we just see the first floor."

"OK," Kino says.

They go up the stairs to the first floor.

Paper Twists

Back up on the first floor, they see a room with kids seated around tables, cutting and pasting paper. This is not really the sort of thing the boys are interested in, but they see that the kids are so engrossed and seem to be enjoying themselves, so they linger by the door.

One of the guides in the room, Toshi, sees them and invites them to join.

"You can sit over there," he says pointing to a table with some empty chairs.

"OK kids, those of you who have just come in, let's warm up!" Toshi calls the kids to attention.

He holds up a paper construction. "Here I have something made from a single sheet of paper. Can you duplicate this construction?"

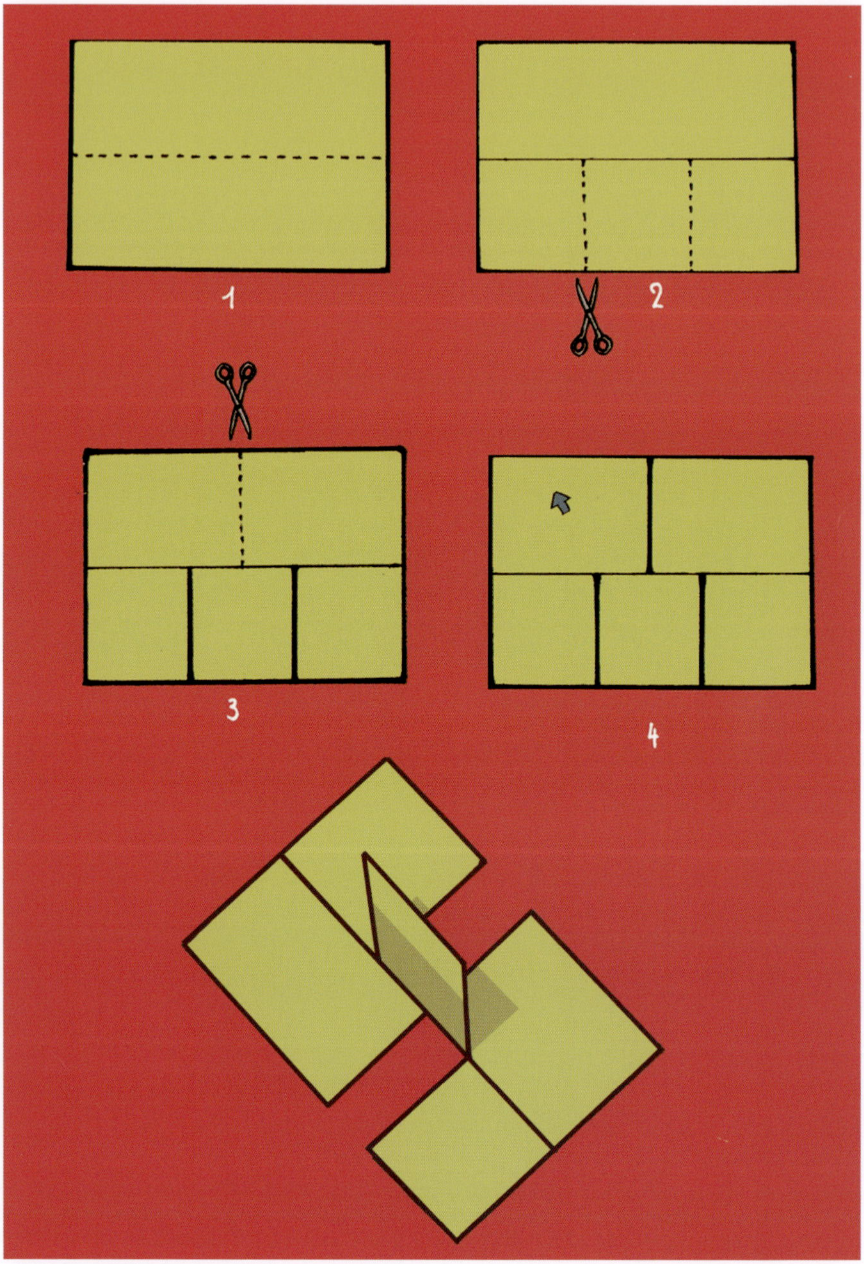
1
2
3
4

"This is challenging," says Kino.

Ichiro and Jai take a hard look at the model. Jai sees the trick right away, Ichiro takes more time but gets it in the end. Like many of the other kids, Kino doesn't know what to do.

"Fold the paper in half lengthwise. Divide one side into three equal parts and cut the dividing lines to the fold. Then divide the other side into two equal parts and again cut to the fold. Twist the half that was last cut by 180°, and you get it," Jai announces.

While Jai is giving instructions, Ichiro is patiently showing Kino how.

Toshi goes around the tables to see how the kids are doing, giving hints, helping out here and there until they all get it.

"You can get a lot of amazing results by twisting paper," he says.

He picks up a strip of paper, one side colored yellow and the other orange. "I will make a half twist on this strip and then paste the ends together."

"Believe it or not, this strip has only one side!" he claims.

"Did he say one side?" Kino asks his friends.

"How's that again?" Ichiro asks Toshi.

"Come and help me demonstrate," Toshi calls Ichiro.

"Mark any starting point on the surface. Then draw a line from the point following the length of the surface," he instructs Ichiro.

Ichiro does as he was told and his line ends up where he started. When he looks at the surface, his line has passed through both the yellow and orange parts of the surface!

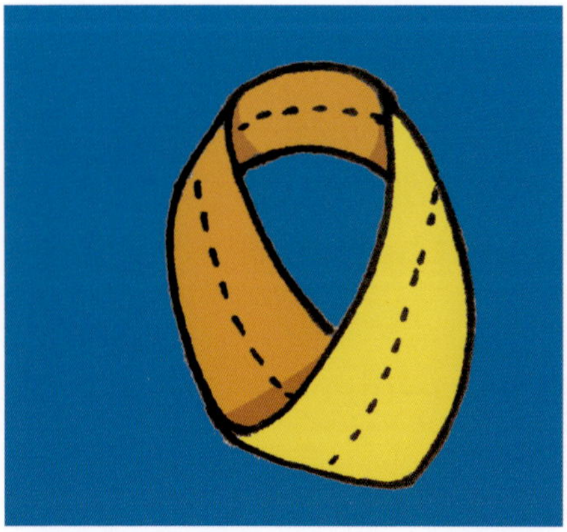

The room is buzzing with excitement at this demonstration.

"This special surface is called a 'Möbius band'," Toshi

says, “This property we just showed makes a Möbius band ideal for use as a conveyor belt. The rate at which it will wear out is half that of the usual circular belt.”

“You will be surprised at what we can do with it,” Toshi declares.

The kids are now wide-eyed with anticipation.

Toshi holds up a prepared Möbius band and proceeds to cut it lengthwise along a line that divides the width into two parts.

“Guess what kind of figure I will get,” he challenges the kids.

“Two Möbius bands each half as wide as the original,” a brave girl answers.

Toshi continues the demonstration and when he holds up the result -- one long band with two half twists, the kids can hardly believe it and they clap.

The guides begin to distribute strips of paper, pencils, paste and scissors.

"Please make your own Möbius bands. Then draw two parallel lines running lengthwise along the band and cut along the lines. See what you get," Toshi tells the kids.

"Maybe we will get one very long band with three twists," Ichiro makes a guess.

The room is quiet as the kids engage in the activity. Soon gasps of surprise and delight are heard as they complete it.

Kino holds up two intertwined Möbius bands, a shorter one with a half twist and a longer one with two half twists.

"This is really weird," Jai says. "What if we make three cuts or four cuts and so on?"

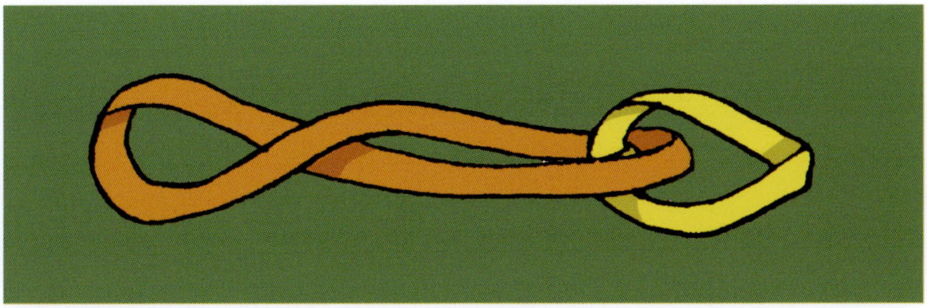

"Try it when you get home. See if you can find a pattern," Toshi says.

Their curiosity aroused, the three boys agree to try this out some time. Jai makes another mental note: Möbius band, three cuts, four cuts, keep going until you find a pattern.

Now Toshi is holding up two circular bands and positioning them so that they are perpendicular to each other. Then he pastes them together. There is a line running along the length of each band dividing the width of the band into two parts.

"I want a volunteer to cut along the lines," Toshi announces, "And the rest of you can guess what the resulting figure will be like."

Ever ready, Kino volunteers.

“Maybe one circular band with width a half of the original bands,” Ichiro says.

“Sounds good to me,” says Jai.

They all watch Kino’s progress. When he finishes, they are all surprised to see a square “picture frame.”

“How did that happen?” Ichiro asks.

Toshi takes an uncut pair of perpendicular circular bands.

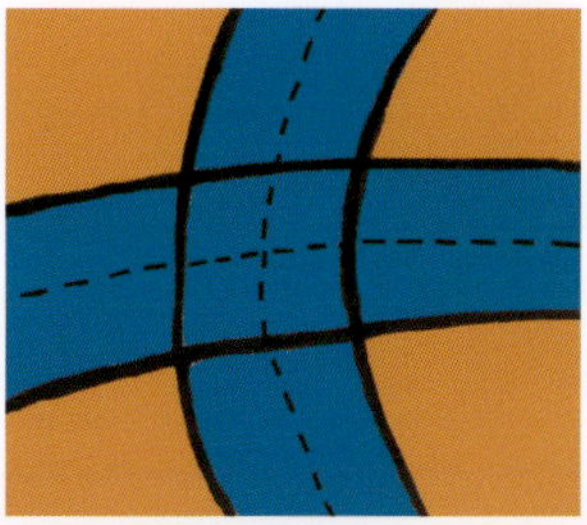

"The corners of the frame will come from this intersection," he says.

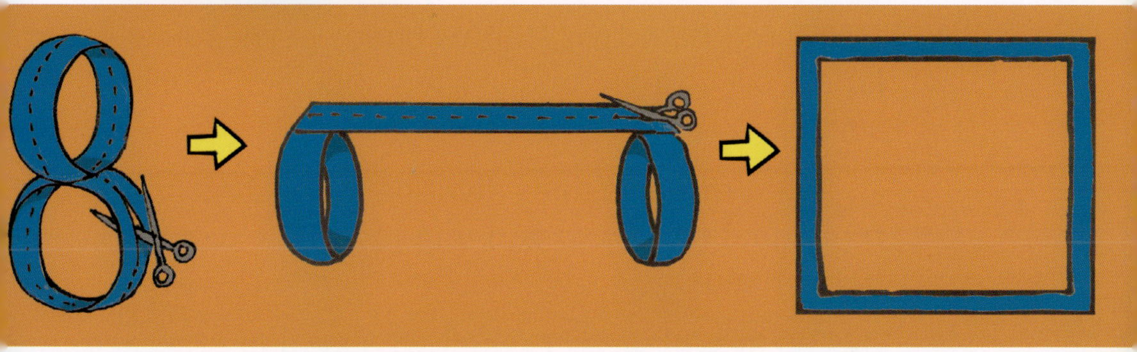

He then demonstrates the cutting again, proceeding in a certain way.

Jai and Ichiro are still discussing this result when Toshi calls their attention to something else.

"Here I have two Möbius bands perpendicular to each other and pasted together," Toshi says.

Again they see lines dividing the widths into two equal parts and running along the lengths of the bands.

"What will we get when we cut along these lines?" he asks.

The kids are overwhelmed. After the previous demonstrations, they don't know what to think, but they are eager to start the activity.

The guides distribute models, like Toshi is holding, to each kid. They quickly begin, curious to find out what they will get.

Ichiro ends up with two intertwined bands that look like hearts, but Kino and Jai end up with two separate bands that also look like hearts.

The boys, as well as the other kids, are amazed and puzzled at the same time.

Kino asks the question in their minds. “How come some of us get separate hearts while the others get intertwined hearts?”

"Very good question," Toshi says. "You are thinking like a mathematician now. It takes observation, good questions, and experimentation."

"You will discover for yourselves." Toshi says.

He divides the kids into two groups. He tells the first group to make two Möbius bands, each twisted in the same direction, then to paste them perpendicularly to one another and cut as before. The second group is to make two Möbius bands, one twisted in one direction and the other in the opposite direction, then to paste and cut like the other group.

After several minues, Toshi asks: "Which group gets the intertwined hearts?"

"Group 2," the kids chorus.

Toshi gives a recap: "To get intertwined hearts, twist the Möbius bands in opposite directions. To get separate hearts, twist in the same direction."

The boys are mesmerized. "Is there more?" they wonder.

I will leave you with an exercise to do at home." The guides give out sheets of paper on which the following are written:

Start with a rather large paper cross. Call the horizontal portion A and the vertical portion B. Draw a line along the length of A that bisects its width.

Draw two lines along the length of B that trisect its width. Join the ends of A without twisting. Give one end of B a half-twist and join it to the other end.

First cut the lines that trisect B, then the lines that bisect A.

Surprise!

Toshi moves on to greet another group of kids.

Fold & Cut

Another guide comes over.

"I'm Yuji," he says. "With Toshi you got some pretty amazing results by twisting paper. Now I'm going to show you some equally amazing results you can get by folding paper."

Yuji holds up what looks like a 4 x 4 paper chessboard. "I will fold this in such a way that one straight scissor cut will make all the black squares fall out."

He folds the chessboard several times, then cuts, and sure enough, the black squares fall out.

The kids clap in appreciation. The boys are wide-eyed with astonishment.

"For starters, I'll give you some simpler figures to cut out," he says.

The guides hand out pieces of paper with one black square.

"Fold the paper so that the black square will fall out with one straight cut," are the instructions Yuji gives.

The boys find this exercise easy, so they are all encouraged. Here is the solution.

"Let's discuss your solution," Yuji says. "What was your strategy?"

"Align all the edges on the boundary of the square," Kino volunteers.

"Did you notice that in order to align the edges, you had to bisect the angles?" Yuji asks.

The boys look back on their solutions.

"We bisected twice," Ichiro notes.

"Bisection is usually part of the strategy," Yuji states. "There is another detail that you must take care of. The boundary lines must align all the way across the place where you plan to make the cut."

With these hints, he asks them to try another exercise. The guides hand out pieces of paper with a triangle.

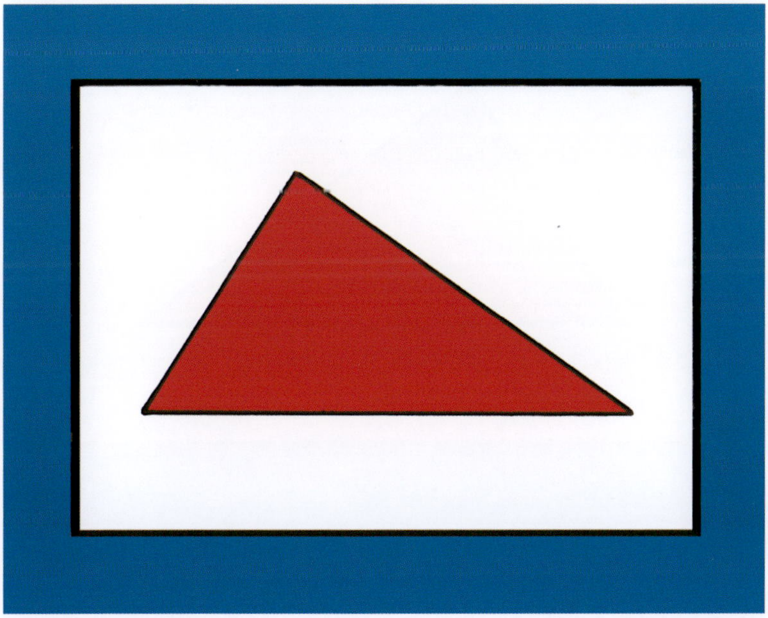

The boys wrestle with the problem. It is not as simple as the first since the triangle is neither equilateral nor isosceles.

"It's not so easy to align the lines this time," Kino comments.

"Maybe we should try bisecting angles like Yuji said," Ichiro suggests.

He folds along the angle bisectors and immediately the paper yields the solution naturally.

"Look what I got," he tells his friends.

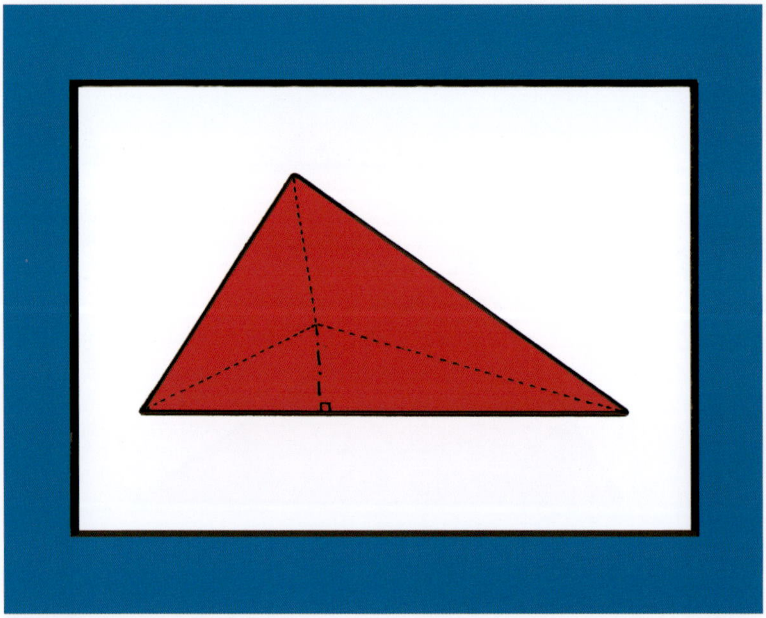

They examine his paper and notice that a fourth fold appears along the perpendicular from the point of intersection of the angle bisectors to one side of the triangle.

"That's cool," Jai says. "Would it work if the fourth fold is made along the perpendicular to another side?"

They try this out using the other two sides, so they discover that the folding is not unique.

Yuji observing them says: "You guys have the right attitude for mathematics, you are asking good questions and pursuing the alternatives."

They are pleased with the compliment.

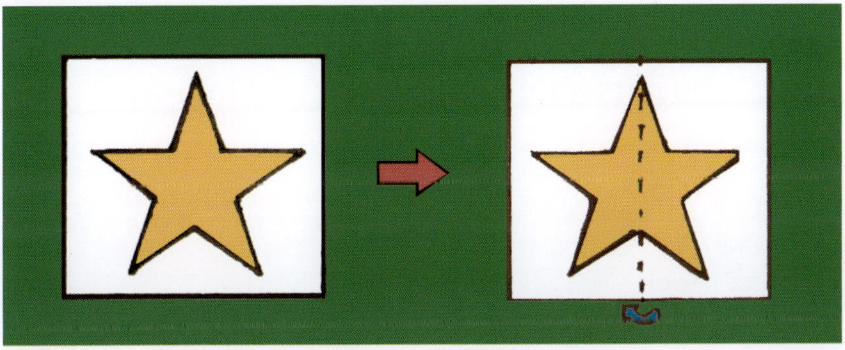

"Try this out," Yuji says handing them three pieces of paper with a five-pointed star.

The boys decide to do this exercise together.

"OK, anyone who has any ideas, speak up," Ichiro tells his friends.

"I think we should fold it by bisecting the angle at the top so that the lines from both sides align," Jai proposes.

They stare at the folded sheet for a while.

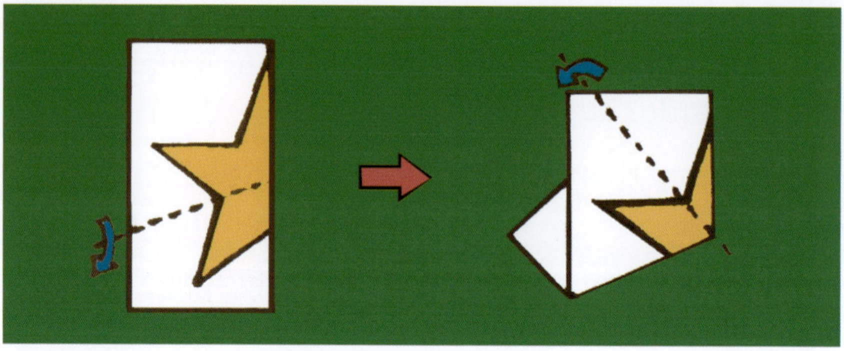

“I know, I know,” Ichiro says excitedly. “We can put the two points together by folding this way.”

He gets hold of the paper and folds it.

“Then we can align the three lines together using two folds,” Jai continues.

Kino cuts their model and the star separates from the rest of the paper.

"We're done." They show their work to Yuji.

"Good work!" he says.

The other kids look at them with a mixture of admiration and envy, since they are way ahead of the rest.

"Ready for another exercise?" Yuji asks.

"Yes!" they say.

He gives them a figure which he says is an old Japanese crest.

They take more time with this but they finally get it done and it gives them a great feeling of satisfaction.

Here is their solution.

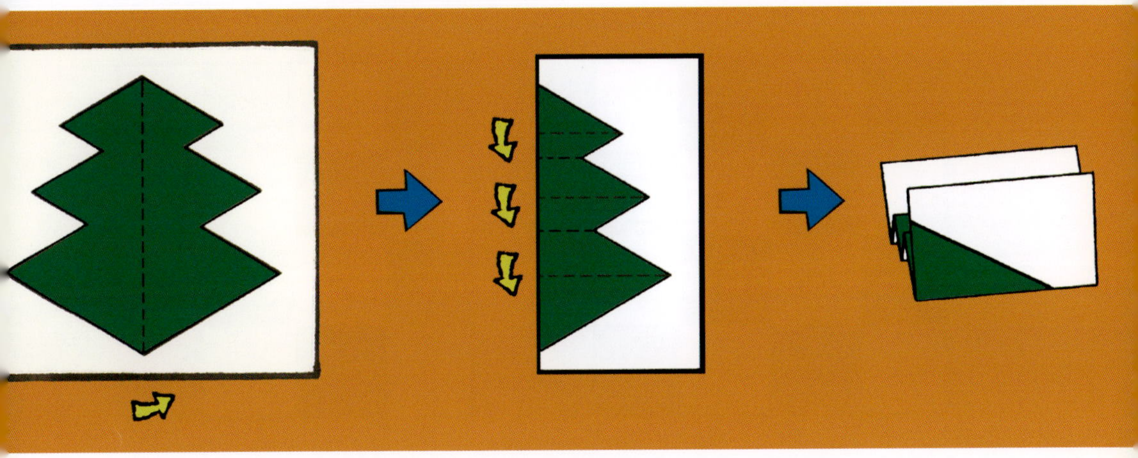

Yuji gives the boys a problem to take home. He hands them a sheet of paper with a trapezoid.

When all the other kids are done with the star, Yuji says: "Now that you have some experience with this kind of problem, I will show you how to fold the 4x4 chessboard that I cut for you at the start."

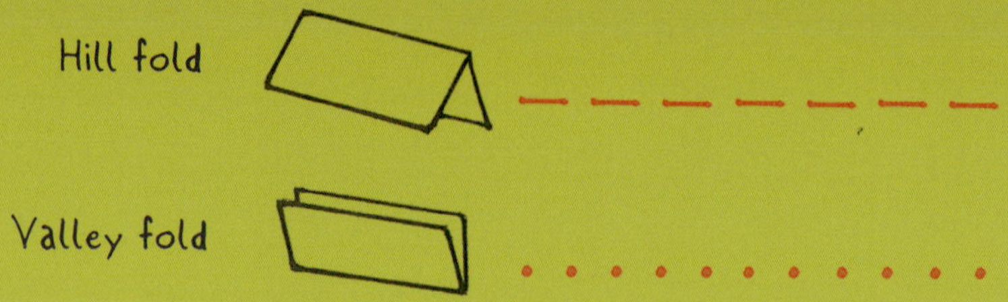

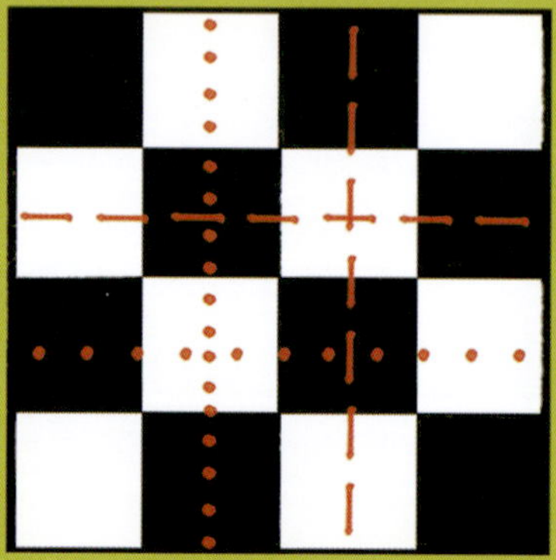

1

Location of the folds

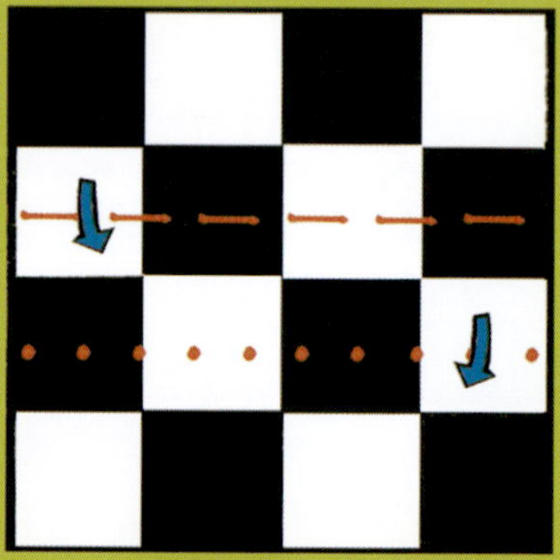

2

5

He goes through the process slowly step by step and when he is done, a single cut releases the black squares.

"Will that work for a bigger chessboard?" Jai asks.

"Only if the number of squares on each side is even," Yuji answers. "If it's odd, you have to make some adjustment."

3

4

6

Cut along a line slightly to the right of the middle.

7

"Great party trick!" Kino says.

Jai files another mental note: try a chessboard with three squares on each side.

They appreciate Yuji's encouragement and go over to thank him.

Jigsaws from Tetrahedrons

The boys sense a commotion in one of the rooms. They come across their first guide Keiko.

"You are in luck!" she says, "Professor Yamaaki is here for a visit. He is the one who collected all the models in Wonderland. You can watch him demonstrate models in the next room."

They are able to find a place close to the front among the gathering crowd of kids. Everyone's attention is focused on a man with unruly salt and pepper hair, a beard like a Chinese sage, glasses and a colorful headband.

He seems friendly. He is joking around with some of the kids.

"I have many things to show you today," he says.

"He looks familiar," Ichiro whispers to his friends, "I think I've seen him before, but I can't remember where."

"Do you know Escher?" the professor asks.

Most of the kids shake their heads, but Ichiro is nodding his. He raises his hands and says

"He's a famous artist."

"How do you know?" Kino asks in private.

"My Mom showed me a book on his works," Ichiro responds. Ichiro was fascinated by the way the figures merged and transformed in Escher's etchings.

"That's right," Professor Yamaaki says.

He shows some copies of Escher's work.

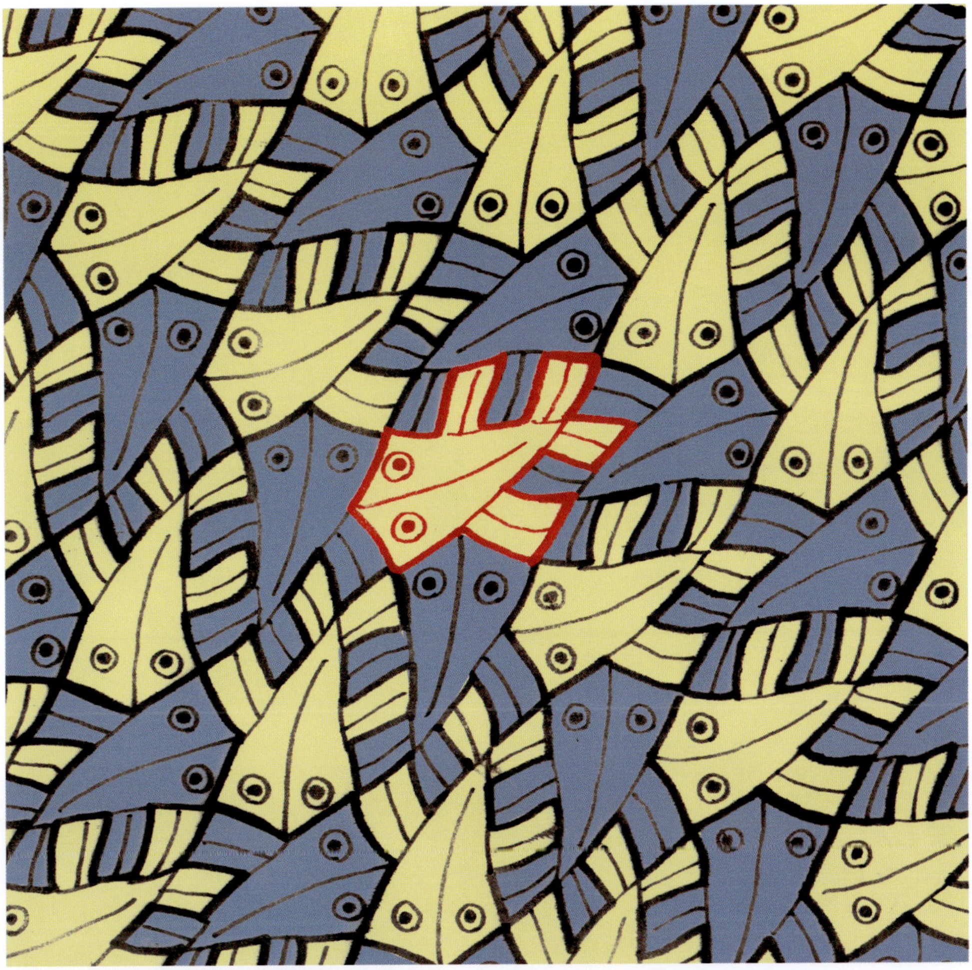

M.C. Escher's "System VI^{B}" © 2008 The M.C. Escher Company-Holland. All rights reserved. www.mcescher.com

"These works have a pattern that is repeated continuously to cover the plane. Notice there are no gaps or overlaps," he continues.

M.C. Escher's "System with Elements from IV and V" © 2008 The M.C. Escher Company-Holland. All rights reserved. www.mcescher.com

"This kind of repeating pattern that covers the plane is called a 'tiling'," he says.[5]

5. Strictly speaking, tilings do not use mirror images of the pattern but Escher's tilings sometimes do.

M.C. Escher's "System X^{E}" © 2008 The M.C. Escher Company-Holland. All rights reserved. www.mcescher.com

"It's just like a jigsaw puzzle where all the pieces have the same shape," Jai volunteers.

"A very good observation," Professor Yamaaki says.

"Today, I will turn you all into artists like Escher," he proclaims.

He holds up a paper model of a solid. "See this solid, it has four sides which are all equilateral triangles of the same size. It is called a regular tetrahedron."

There are a lot of tetrahedrons lying on a table behind the professor. Soon the guides are handing a tetrahedron to each kid together with a cutter, a pair of scissors, a large sheet of paper and some paste.

The boys get brightly colored tetrahedrons. They turn them around checking out the triangles.

"I want you to cut the surface of the tetrahedron carefully, so that it remains in one piece and so that you can lay the piece flat on a table," are the instructions that Professor Yamaaki gives.

"How do you do that?" Kino asks.

Professor Yamaaki cuts a tetrahedron to demonstrate. “The trick is to make a continuous cut that passes through each vertex of the tetrahedron.”

“You will find that your tetrahedron has several layers, each of a different color,” he continues.

When he finishes cutting, he displays the pieces on a board and shows how they all fit together like a jigsaw puzzle.

"Cut in any way you want, just make sure that you end up with just one piece and you can lay it flat," Professor Yamaaki repeats. "Then paste your work on the sheets of paper."

Jai's snails

Kino's sailboats

The kids engage in the activity with gusto. Ichiro wants to create a robot design. Jai is trying out something very convoluted. Kino is happy with a simple cut that he can lay flat.

Ichiro's robots

Professor Yamaaki goes around to show off some of the best patterns. The kids clap when they see Ichiro's robots.

"This is one of my newest discoveries," the professor announces. "No matter how you cut the surface of a regular tetrahedron, as long as it remains in one piece and it can be lain flat, the resulting pattern will tile the plane."

"That means we can make all kinds of designs!" Ichiro says.

"But we have to make the tetrahedrons first," Kino notes.

Jai is already thinking of the designs he will try out at home.

Single & Double Duty Solids

The kids follow the Professor to another room. On the floor are large wooden blocks. Some kids, already in the room, try to fit them together like three-dimensional jig-saws.

"In the other room, I showed you a way to tile the plane. Just as you can tile the plane, you can also fill space," Professor Yamaaki says.

"Look at the kids over there, that's what they are doing."

He goes over to one group of kids and picks up one of the wooden blocks.

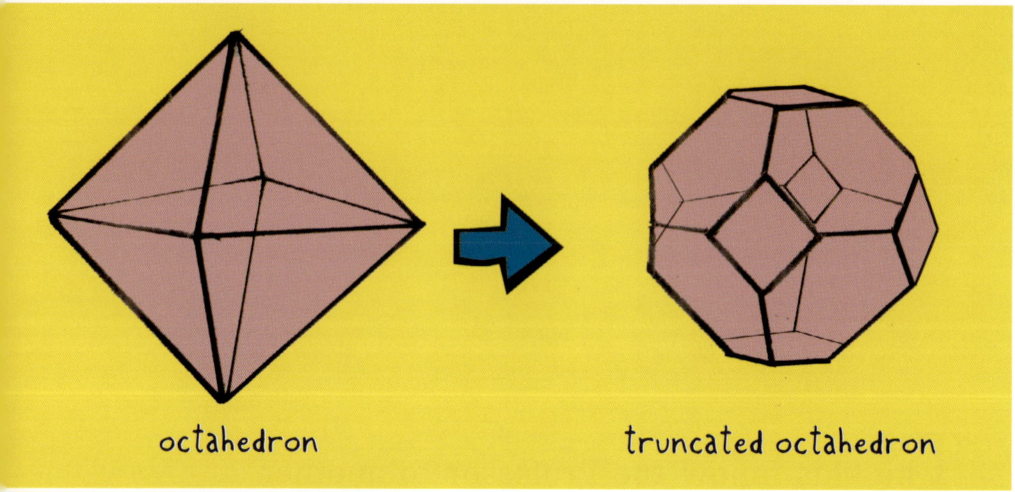

"This is a truncated octahedron. It has 14 faces consisting of eight regular hexagons and six squares." he says. "You can make it from an octahedron by cutting off its six vertices."

He shows an octahedron and makes some chopping motions as if he is cutting off its vertices.

"The truncated octahedron is a space-filler. That means you can fill up three-dimensional space so that there are no gaps or overlaps using only truncated octahedrons."

He picks up another kind of solid.

"This is a rhombic dodecahedron. It has 12 faces, each of them a rhombus. It is another space-filler. There are many solids which are space-fillers."

He moves to another group of solids. “These are also space-fillers but they are very special; not only do they fill space, but there is also a way to cut their surfaces and lay them flat so that the resulting plane figures tile the plane. Let’s call them ‘double-duty solids’.”

“Nice name, easy to remember,” Ichiro says. “Double-duty because they do two things -- fill space while some surface cut tiles the plane.”

“Then the ones that only fill space should be called ‘single-duty solids’,” Kino remarks.

“I’ll go with that,” the professor says after overhearing them. “Let’s call space-filling solids single-duty solids.”

“One double-duty solid is very similar to the tetrapak containers sometimes used for milk and juice. It is a tetrahedron whose sides are isosceles triangles with edges in the ratio 2:$\sqrt{3}$: $\sqrt{3}$.”

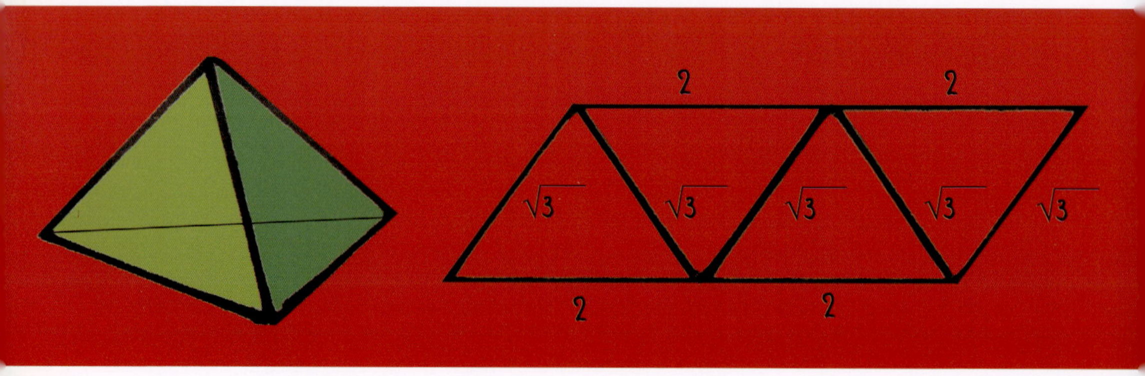

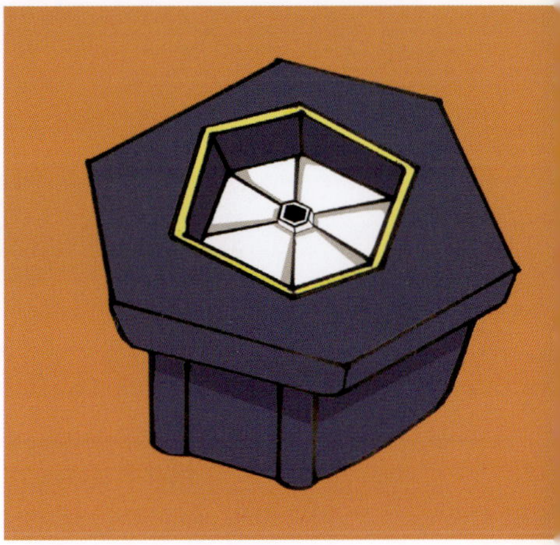

"Look, you can stack them to form right triangular cylinders. Then these can be put together to form a hexagonal cylinder. And this is precisely how tetrapak containers are transported."

He takes a paper tetrapak and cuts the surface and lays it flat.

“Certainly, this can tile the plane,” he continues, referring to the cut surface.

“Cubes, of course, are double-duty solids. In fact if you confine the surface cuts to edges, all the plane figures you get from a cube with tile the plane.”

He leads them to a poster showing such tilings.

Tiling the Plane with Edge-Developments of a Cube

An edge-development of a polygon is a connected plane figure obtained by cutting the surface of the polygon along certain edges.

All Possible Edge-Developments of a Cube

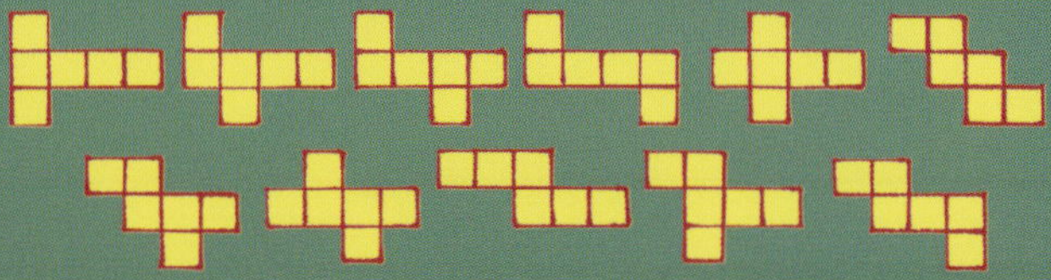

Tilings of the Plane Using the Edge-Developments of a Cube

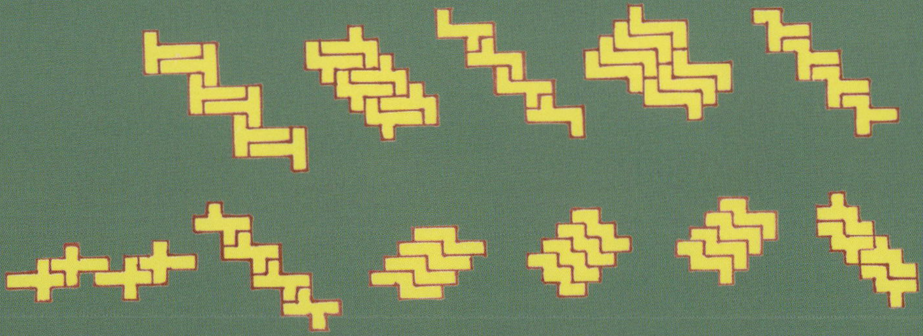

He picks up another wooden model.

"There are infinitely many double-duty solids, but among the known ones, this dodecahedron has the largest number of faces. It's like a cube with two right pyramids attached to opposite faces."

Again he cuts a paper model of the dodecahedron, lays it flat and shows that it tiles the plane.

"Maybe I can find an undiscovered double-duty solid," Jai muses.

While the professor goes around the room talking to some of the kids, the boys go to see the other single and double-duty solids displayed in the room.

On a table are three congruent skewed pyramids. Each has a square base and four triangular sides. The boys see a note on the table

These three pieces can be assembled to form a cube.

"Let's try it," Kino says.

The boys attempt to form a cube.

Ichiro has an idea. "Since the bases are square, they must be the sides of the cube."

With this in mind, they are able to form the cube.

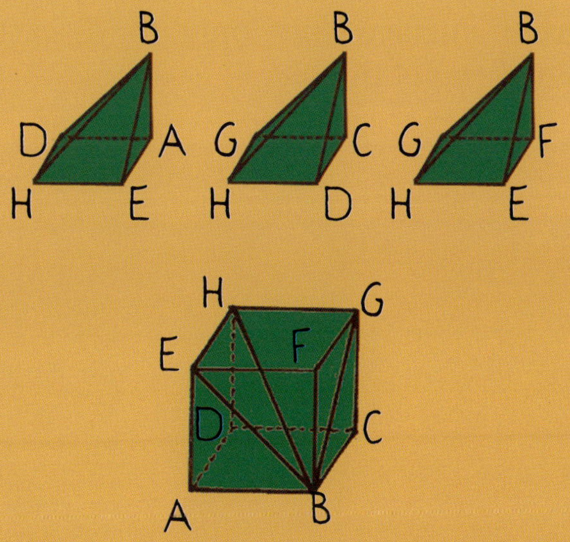

"Since congruent pieces of this solid forms a cube, it must be a single-duty solid," Jai says.

"What's a single-duty solid?" a guide named Masao asks them.

"Oh, that's a name we invented for a space-filler," Kino answers.

"In fact, it's more than that," Masao says.

"It also has a surface cut that tiles the plane. Can you find the surface cut?"

He gives them paper and pencils. “Trace the faces of the solid and try to find the tile,” Masao challenges them.

The problem is not so easy, they draw many different surface cuts only to discard them. Finally they get the tile.

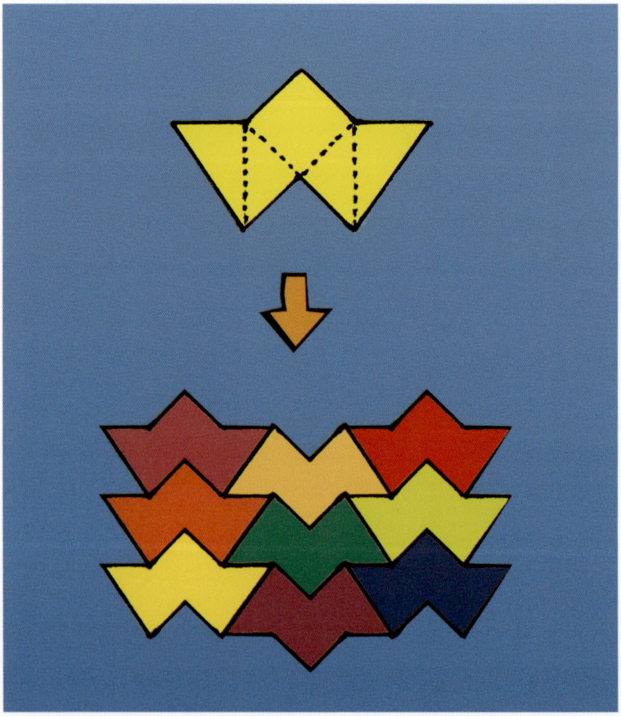

“Good job!” Masao says when he checks on them.

On another table are many congruent solids with an unusual shape. Each has six faces, each face a quadrilateral.

"That solid is quite special," Masao says. "Many pieces of it were sent to Professor Yamaaki by a carpenter who had seen one of his TV lectures on space-filling solids."

"What's the solid called?" Kino asks.

"It has no name as yet. Want to give it a name?" Masao asks the boys.

The boys get into a huddle and discuss the relevant words 'carpenter', 'six-sided', 'quadrilaterals' and 'polyhedron'. They choose certain letters from the four words:

*c*arpenter *s*ix-sides *quad*rilaterals polyhed*ron*

"We'll call it the '*c-squadron*'," they tell Masao.

"We'll call it that then," Masao says.

"So is it a single-duty solid?" Ichiro asks.

"Yes it fills space," Masao responds.

He leads them to solids assembled from many c-squadrons. "See, you can construct other space-fillers from them."

They see a tetrapak, a truncated octahedron, a rhombic dodecahedron.

“You can assemble a tetrapak using four c-squadrons, a truncated octahedron using twenty-four, and a rhombic dedecahedron using ninety-six,” Masao adds.

The boys spend some time trying to replicate the constructions.

Jai is thinking: “When I discover a new double-duty solid, I will call it ‘jaihedron’.”

Reversible Solids

OPEN!
OPEN!

Professor Yamaaki moves to another table. On it is a triangular prism with thc word "open" painted on it.

He picks up two identical solids from the table.

"These are truncated octahedrons," he says.

He then holds up a plastic box.

"I want to pack the two truncated octahedrons into this box."

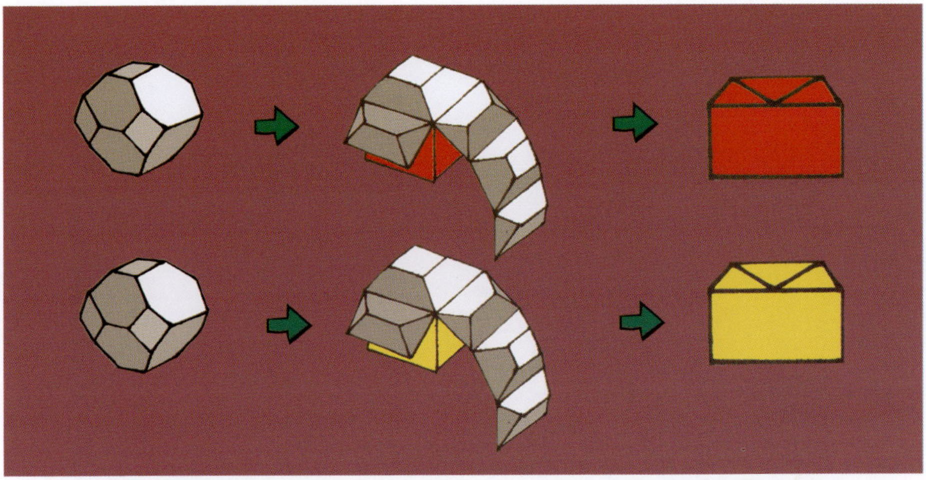

To most of the kids this seems like an impossible task, judging from the size of the truncated octahedrons and the size of the box. And of course Professor Yamaaki's initial attempts to put them in fails.

Then, with a few quick motions, Prof. Yamaaki turns the solids inside out to get two identical bricks that slide easily into the box.

The three boys are impressed. Everyone claps!

"Sometimes, to solve a problem, you have to change your way of thinking," the professor says.

Another mental note for Jai: There are many different ways of looking at a problem.

The professor then picks up a solid painted to look like a fox.

“This solid is a rhombic dodecahedron, but for now, we will pretend that it is a fox,” he says.

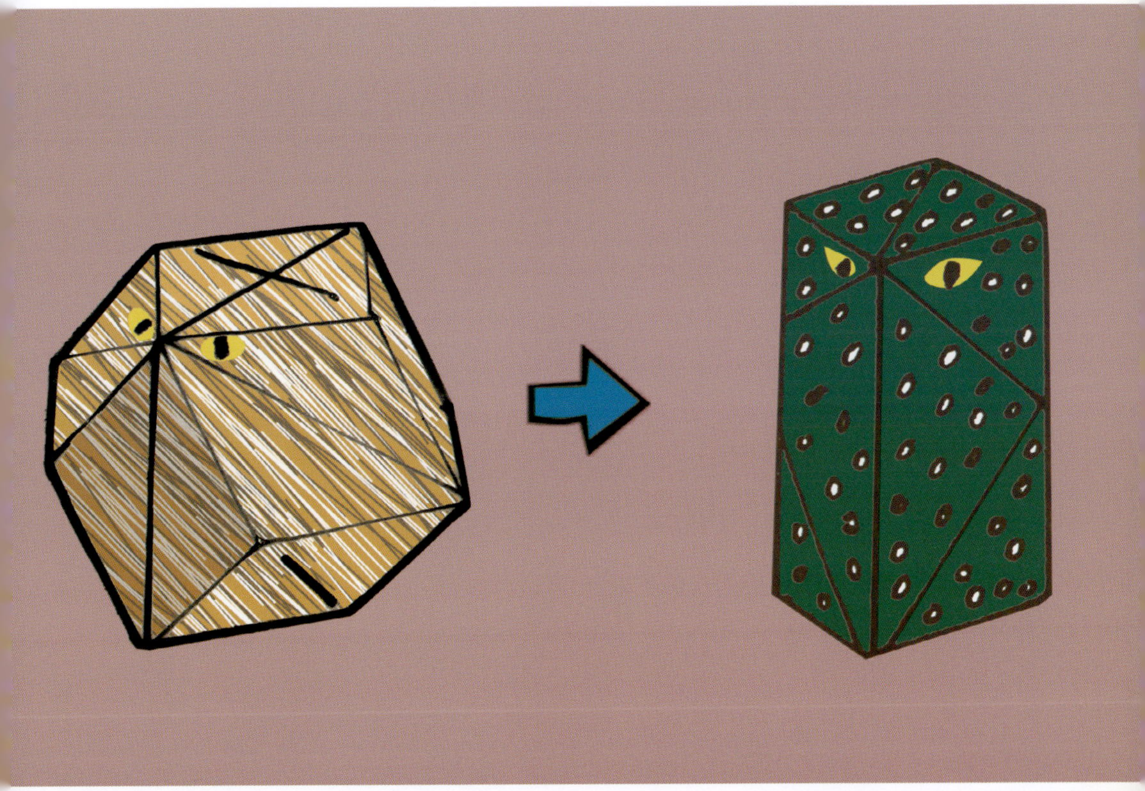

He pulls a string and the solid is turned inside out into a rectangular solid painted to look like a snake.

“The snake ate the fox!” he says in a voice filled with mischief.

“Aah...” the kids utter a collective sigh.

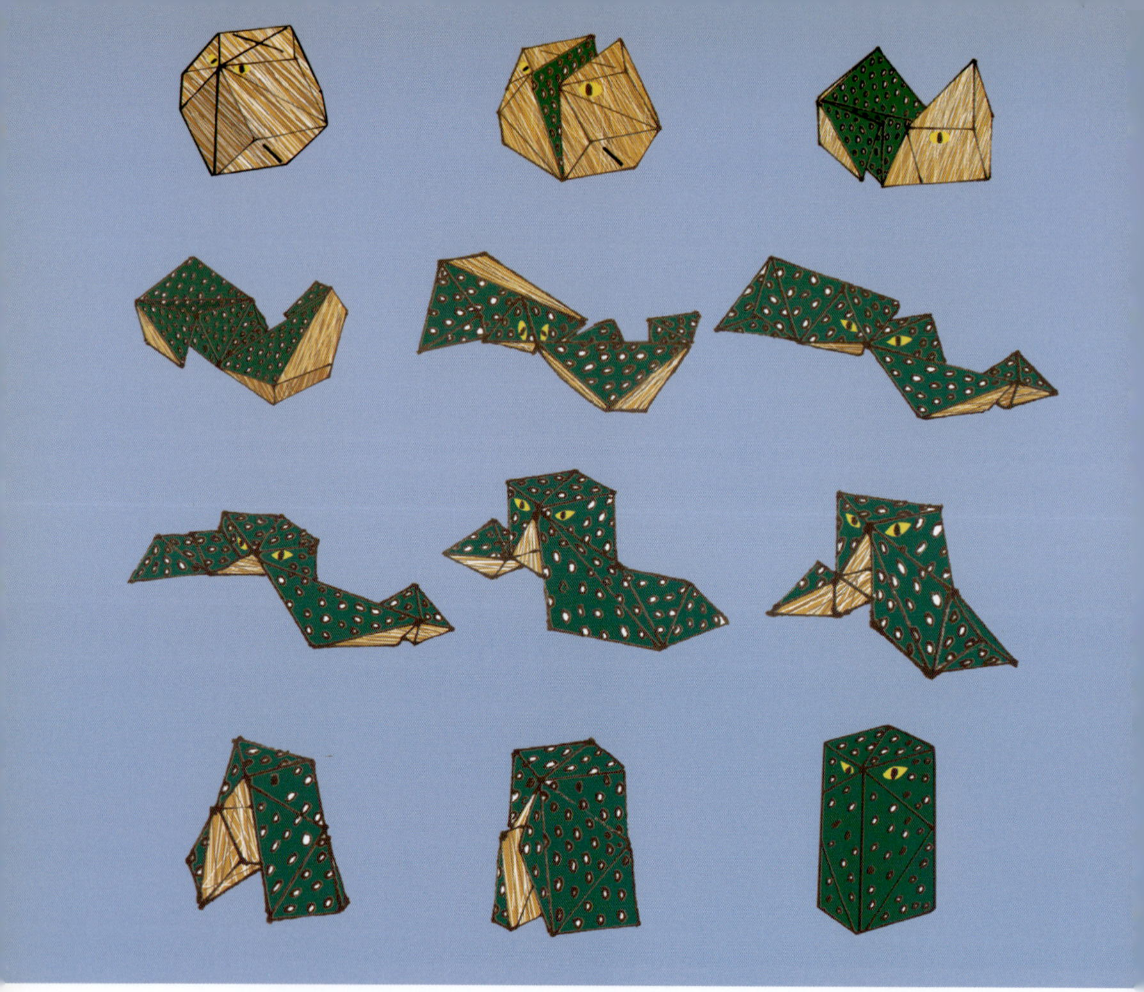

He proceeds to do the previous trick slowly, step-by-step, so that the children can observe what is happening.

"Some solids can be cut in a special way and turned inside out to form other solids. I call them 'reversible solids'. Sometimes, we can cut and convert them into solids congruent to themselves."

He shows them another solid. Green chameleons are painted on it.

“Here’s another truncated octahedron,” he says.

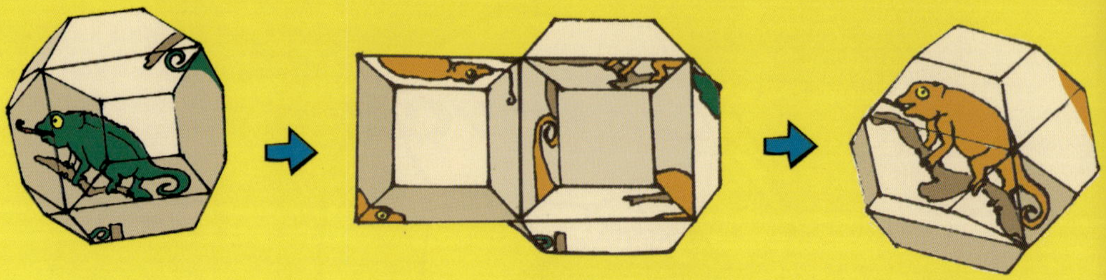

He turns it inside out into another truncated octahedron, this time with orange chameleons painted on it.

"So this is an example of a solid that can be turned inside out to form a congruent solid. I call these reversible solids 'chameleons'," he continues.

Jai has observed something. "So the same solid can sometimes be turned inside out to get two different solids?" he asks Professor Yamaaki.

"A very good observation," the professor says. "We converted the first truncated octahedron into a rectangular solid, while the second truncated octahedron, the chameleon, we turned into another truncated octahedron. Yes, with a different way of cutting, certain solids can be turned into two or more different solids."

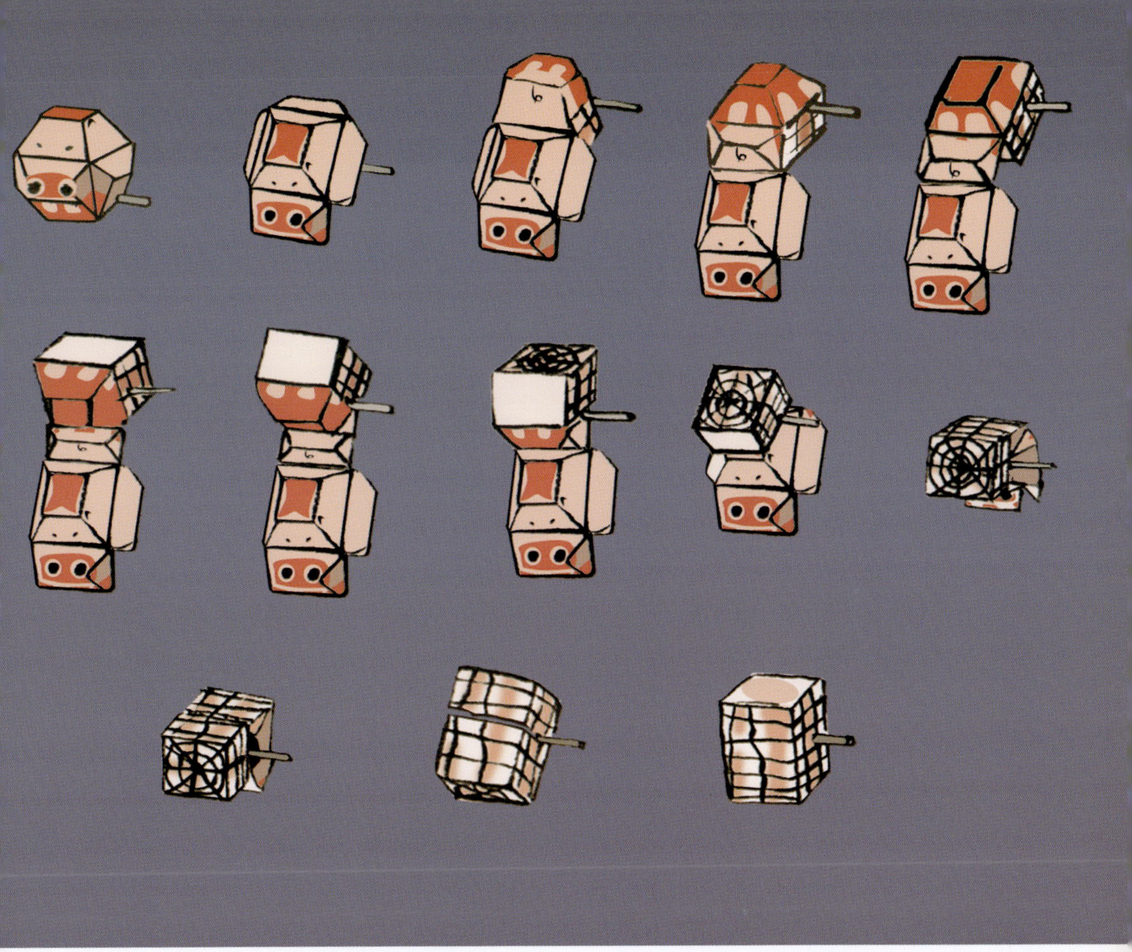

“I have made many solids that can be turned inside out to form other solids. Many of them are displayed in this room.”

He picks up another solid painted to look like a pig. Jai is able to identify the solid as another truncated octahedron.

The pig is attached to a rod. With a twist of the rod, Professor Yamaaki turns the pig inside out and the pig is converted to a slab of ham! All the kids laugh out loud.

OPEN!
OPEN!
OPEN!
OPEN!

The professor picks up the triangular prism with the word “open” and transforms it into a congruent triangular prism with the word “closed”, to indicate the end of his demonstration.

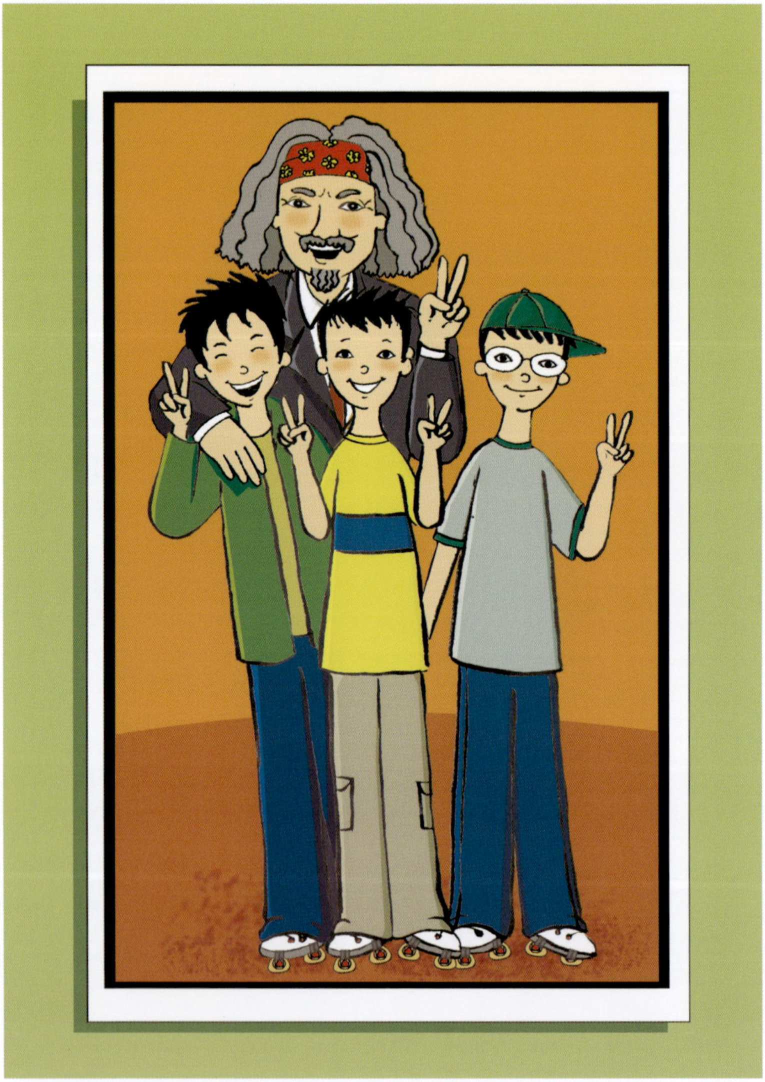

The boys are amused, amazed, agog. They are suffering from sensory overload. They join a line of kids waiting to get Professor Yamaaki's autograph and to have a picture taken with him.

All the Way Home

Wonderland is about to close for the day. The boys have stayed as long as possible. They make their way to the counter to return their skates. They have their take-home problems, paper models and cut-outs.

On their way out, they notice a bank of TV screens. Professor Yamaaki is on all of them and he is demonstrating something different on each screen.

"Oh, that's why he looks so familiar." Ichiro says, "I have seen him on TV."

They meet Keiko again. She shows them a notebook. "This is where our visitors write their comments. Would you like to write something?"

The boys take turns to write.

I didn't like math before, but now I think I could like it. The machine that makes square holes is great! So is the machine that computes GCF and LCM.

Ichiro

Math is fun! But I didn't know before I came here. I like the slides best. Where can we buy the skates?

Kino

Amazing and challenging! I never had to think as hard as I did today and yet I enjoyed it so much. I wonder if I can be a mathematician.

Jai

P.S. We didn't see everything. We'll be back.

The boys are very tired, so they are quiet during most of the ride home. Each one is left to his own thoughts.

M.C. Escher's "Metamorphosis III" © 2008 The M.C. Escher Company-Holland. All rights reserved. www.mcescher.com

Ichiro is thinking that Grandma was right after all. Wonderland was a lot of fun. Math can be fun. The intertwined hearts are for Grandma and the "Escher" is for Mom. Maybe she will hang it beside that Escher print in her room.

Kino is reviewing all the fun things he experienced. He is trying to figure out what he can say about this trip to upstage Kentaro.

Jai is recalling all his mental notes and planning the math experiments he will do at home.

ACKNOWLEDGMENTS

Joseph O'Rourke alerted us to the existence of bicycles with square wheels in certain science museums in the U.S.A. and Stan Wagon confirmed it.

It was in Erik D. Demaine's lecture "Folding and Unfolding Linkages, Paper and Polyhedra" at the Japan Conference on Discrete and Computational Geometry 2000 that we first learned about the fold and cut problem. He graciously allowed the use of some of his crease patterns in Jin Akiyama's TV lectures.

Sebastian von Wuthenau Mayer and Claudia Masferrer Leon invented the wheel and axle mechanism found in the skates featured in this book. The wheels have constant width and the axles are Reuleaux triangles circumscribed by square frames within which they rotate.

Yasuyuki Yamaguchi and Minoru Kanzaki fabricated most of the models in Math Wonderland.

The models for the triangular and pentagonal holes were provided by Yoshiyuki Kawazoe and Ikuro Sato.

I. Sato and H. Nakagawa provided the space-filling solids which we call c-squadrons in the book.

The M.C. Escher Company-Holland granted permission to use the impressions of the Escher prints in this book.

The Hyatt Co. obtained a patent in the U.S.A. in 1921 for a drill that makes square holes. Their drill has a fixed blade and a rotating framework and weighs 200 kg. The drill in Math Wonderland has rotating blades, a flexible axis, a fixed framework and is portable.

In 1889, Francis Galton designed a board called the Galton quincunx to demonstrate the normal distribution. The top of the *pachinko* device found in Chapter 6 is adopted from Galton's board, but the cradle bottom was added to make it easier to tilt the device for the demonstrations and also to return the balls to their compartment after.

REFERENCES

General

Mathematical Art Catalog, Research Institute of Educational Development, Tokai University, 1999

Mathematical Art in ICME 9 Catalog, Research Institute of Educational Development, Tokai University and Chiba Convention Bureau, 2000

Chapter 2

J. Akiyama, *A Book of Mathematical Problems for Someone Special* (in Japanese), Gentosha, Tokyo, 1999

C. Gardner, *Unexpected Hanging and Other Mathematical Diversions*, Simon and Schuster, Inc., New York, 1974

P. Rexford, "Sunk Ships Sail Again in Coins," *The Washington Times*, Sept. 6, 2006

S. von Wuthenau Mayer and C. Masferrer Leon, private communication

http://www.cut-the-knot.org/do_you_know/cwidth.shtml

http://www.engin.swarthmore.edu/~nlaport1/wankel.html

Chapter 3

W. Dunham, *Journey Through Genius*, Penguin Books, New York, 1990

D. H. Shin and S. Singh, Path Generation for Robot Vehicles Using Composite Clothoid Segment, tech report CMU-RI-TR-90-31, Robotics, Carnegie Mellon University, 1990

S. Wagon, *Mathematica in Action*, W. H. Freeman and Company, New York, 1991

D. Wells, *The Penguin Dictionary of Curious and Interesting Geometry*, Penguin Books, London, 1991

http://cage.rug.ac.be/~hs/wheels.html

http://ffden2.phys.uaf.edu/211_fall2002.web.dir/Shawna_Sastomoinen/Clothoid_Loop.htm

http://ffden2.phys.uaf.edu/211_fall2002.web. dir/Shawna_Sastomoinen/Roller_Coasters.htm

http://knuttz.net/hosted_pages/Eejanaika-Roller-Coaster

http://mathworld.wolfram.com/Tautochrone_Problem.html

http://mathworld.wolfram.com/Cycloid.html

http://www.2dcurers.com/spiral/spirale.html

http://www.phy6.org/stargaze/Sclothoid.htm

http://www.teachingtools.com/GoFigure/FlyerCarpets.htm

http://xahlee.org/SpecialPlaneCurves_dir/Epitrochoid_dir/ epitrochoid.html

Math Trek: Riding on Square Wheels by I. Peterson in http://www.sciencenews.org/20040403/mathtrek.asp

Chapter 4

J. Akiyama et al., *Models and Experiments for Junior High School and High School Students* (in Japanese), Suken, Tokyo, 1999

C. B. Boyer, *A History of Mathematics*, John Wiley & Sons, Inc., New York, 1968

I. Moscovich, *Leonardo's Minor and Other Puzzles,* Sterling Publishing Co., Inc., N.Y. 2004

How Did Pythagoras Do It? by E. von Glaserfield in http://www.univie.ac.at/constructivism/EvG/

Chapter 5

http://www.us.es/ewcg04/cosmoakiyama.pdf

Chapter 6

J. Akiyama and N. Torigoe, Teaching probability distributions with a cradle pinball device, Proc. 8th Southeast Asian Conf. on Math. Ed., Ateneo de Manila University,1999, 69-76

http://www.subtangent.com/maths/ig-quincunx.php

Chapter 9

G. David and C. Tomei, The problem of the calissons, *American Math. Monthly*, 96, 5(1989) 429-431

Chapter 10

C. Clawson, *Mathematical Sorcery,* Plenum Trade, New York, 1999

R. Cooke, *The History of Mathematics*, John Wiley & Sons, Inc., New York, 1997

S.A. Garfunkel et al., For All Practical Purposes: Introduction to Contemporary Mathematics (4th ed.), W. H. Freeman and Company, New York, 1997

D. Hilbert and S. Cohn-Vossen, *Auschailiche Geometri*e, Verlag von Julius Springer, Berlin, 1932

M. Sobel and N. Lerner, *Algebra and Trigonometry* (5th ed.), Prentice-Hall, Englewood Cliffs, N.J., 1995

http://www.us.es/ecwg04/cosmoakiyama.pdf

Chapter 11

M. Gardner, *Are Universes Thicker than Blackberries?*, W.W. Norton and Company, New York, 2003

K. Tsubota, Pleasant Math, *Hands on Math* (in Japanese), Kyoiku Publishing Co. Ltd., Tokyo, 2004

Chapter 12

E. Demaine and M.Demaine, Fold and Cut Magic in *Tribute to a Mathemagician*, B. Cipra, et al. (eds.), A.K. Peters Limited, Wellesley, MA., 2005

E. Demaine, Folding and unfolding linkages, paper, polyhedra, *Discrete and Comput. Geom.*, LCNS 2098, J. Akiyama, M. Kano and M. Urabe (eds.), Springer-Verlag (2001), 1-30

Chapter 13

J. Akiyama, Tilemakers and semi-tilemakers, *American Math. Monthly*, 114 (2007), 602-609

J. Akiyama, You Can Be an Artist like Escher: Art from Tilings of Plane, Research Institute of Educational Development, Tokai University, Tokyo, 2006

D. Schattschneider, *M.C. Escher: Visions of Symmetry,* W.H.Freeman and Company, New York, 1990

Chapter 14

J. Akiyama and G. Nakamura, A lesson on double-packable solids, *Teaching Mathematics and its Applications*, 18 No.1, Oxford University Press (1999) 30 - 33

I. Sato and H. Nakagawa, *Wooden Polyhedra* (in Japanese), NPO Center for Collaborative Interdisciplinary Sciences, Sendai, 2006

Chapter 15

J. Akiyama and G. Nakamura, Dudeney dissections of polygons and polyhedrons, a survey, *Discrete and Comput. Geom.*, LCNS 2098, J.Akiyama, M.Kano and M. Urabe (eds.), Springer-Verlag (2001) 1 - 30